the ARTIFICIAL *Intelligence* handbook

Business Applications in Accounting, Banking, Finance, Management, Marketing

JOEL G. SIEGEL
Queens College of the City of New York

JAE K. SHIM
California State University, Long Beach

JOHN P. WALKER
Queens College of the City of New York

ANIQUE A. QURESHI
Queens College of the City of New York

SUSANNE O'CALLAGHAN
Pace University

PAUL KOKU
Florida Atlantic University

THOMSON

SOUTH-WESTERN

Australia · Canada · Mexico · Singapore · Spain · United Kingdom · United States

The Artificial Intelligence Handbook: Business Applications in Accounting, Banking, Finance, Management, and Marketing
Joel G. Siegel, Jae K. Shim, John P. Walker, Anique Ahmed Qureshi, Susanne O'Callaghan, Paul Koku

Editor-in-Chief:
Jack Calhoun

Vice President/ Executive Publisher:
Dave Shaut

Acquisitions Editor:
Steve Momper

Channel Manager, Professional:
Mark Linton

Production Editor:
Darrell Frye

Production Manager:
Tricia Matthews Boies

Manufacturing Coordinator:
Charlene Taylor

Compositor:
Edgewater Editorial Services

Editorial Associate:
Michael Jeffers

Production Associate:
Barbara Evans

Printer:
Phoenix Book Technology
Hagerstown, MD

Cover Designer:
Chris Miller

COPYRIGHT © 2003 by South-Western, a division of Thomson Learning. Thomson Learning™ is a trademark used herein under license.

Printed in the United States of America
1 2 3 4 5 05 04 03 02

For more information contact South-Western,
5191 Natorp Boulevard,
Mason, Ohio 45040.
Or you can visit our Internet site at:
http://www.swcollege.com

ALL RIGHTS RESERVED.
No part of this work covered by the copyright hereon may be reproduced or used in any form or by any means—graphic, electronic, or mechanical, including photocopying, recording, taping, Web distribution or information storage and retrieval systems—without the written permission of the publisher.

For permission to use material from this text or product, contact us by
Tel (800) 730-2214
Fax (800) 730-2215
http://www.thomsonrights.com

ISBN: 0-538-72697-0

Dedication

The management and staff at the Cherrywood Deli in Wantagh, New York. The most competent, hard working, and nicest people.

Acknowledgments

We express our deep appreciation to Barbara Evans for her outstanding editorial work.

Special thanks goes to Roberta M. Siegel for contributing her expertise on artificial intelligence applications in business. We very much appreciate Roberta's advice and suggestions.

Table of Contents

About the Authors xi
What This Book Will Do For You xv

Chapter 1 Introduction 1
   Conventional versus Artificial Intelligence Applications 2
   Expert System Shells and Products 2
   Business Applications of Artificial Intelligence 3
   What is a Machine 6
   Genetic Algorithms 9
   Turing's Test 10
   Eliza 12
   Processing 13
   Branches of AI 14
   Programming for AI 14
   Backward Chaining 15
   Bots and Agents 15
   Fuzzy Logic 16
   The Java Expert System Shell 17
   Instant Tea 17
   Appendix 1-A: Can Computers Think? 20
   Appendix 1-B: Questions for Turing Test 22

Chapter 2 Artificial Intelligence and Expert Systems 23
    Expert Systems in Detail 24
    How Expert Systems Work 25
    ES Tools 26
    Applications of Expert Systems 28
    Benefits and Disadvantages of Expert Systems 36
    Other Applications 37
    Intelligent Computer-Aided Instruction (ICAI) 40

Chapter 3 Neural Networks 41
    Business Neural Network Applications .42
    Insolvency Prediction 43
    References 47

Chapter 4 Artificial Intelligence in Business Finance 49
    Banking: Granting and Monitoring Credit 50
    Insurance: Underwriting, Claims Processing, and Reserving 51
    Portfolio Management 53
    Advice on Trading 56
    Global Financial Markets 57
    Neural Networks in Business Finance 60

Chapter 5 Artificial Intelligence in Accounting 93
    Expert System Applications 93
    Neural Networks in Accounting 107

Chapter 6 Legal Applications 135
    Finder 136
    Shyster 137
    Applied Sentencing System 138
    Legal Information Retrieval System 139
    The Nervous Shock Advisor 139
    Negligence 140
    Domestic Violence 140
    The Inheritance Advisor 140
    Appendix 6-A: Sample Consultation Hypothetical for NSA . . .142

Table of Contents ix

Chapter 7   Artificial Intelligence in Marketing 149
    Using AI at the Design Stage 151
    Selling 153
    Marketing Services 159
    Forecasting 160
    Inventory Management 164
    Trend Analysis 165
    Communication 165
    Making Decisions in Uncertainty 166
    What's Next? 167
    References 168

Chapter 8   Expert Systems and Neural Networks in
                Manufacturing 171
    Example One: Expert Systems Integration Drives
        Van Conversions 172
    Example Two: General Electric's Expert System Mimics
        Human Repair Experts 173
    Production and Operations 174
    Business Process Reengineering 176
    Robotics 177
    Neural Networks in Manufacturing 177

Appendix I: Artificial Intelligence Software 181

Glossary 189

Index 223

About the Authors

JOEL G. SIEGEL, Ph.D., CPA, is a self-employed certified public accountant and professor of accounting and finance at Queens College of the City University of New York. He was previously employed by Coopers and Lybrand, CPAs, and Arthur Andersen, CPAs. Dr. Siegel has been a consultant in accounting and finance to many organizations, including Citicorp, International Telephone and Telegraph, United Technologies, and the American Institute of CPAs. Dr. Siegel has written 65 books and more than 200 articles on accounting and financial topics. His books have been published by Glenlake Publishing Company, Ltd., South-Western, Prentice-Hall, Richard Irwin, Probus, Macmillan, McGraw-Hill, HarperCollins, John Wiley, International Publishing, Barrons, American Management Association, and the American Institute of CPAs. His articles have been published in many accounting and financial journals including *Financial Executive, The Financial Analysts Journal, The CPA Journal, Practical Accountant,* and *The National Public Accountant.* In 1972, he was the recipient of the Outstanding Educator of America Award. Dr. Siegel is listed in *Who's Where among Writers* and *Who's Who in the World.* He served as Chairperson of the National Oversight Board.

JAE K. SHIM, Ph.D., is an accounting consultant to several companies and professor of accounting at California State University, Long Beach. He received his Ph.D. degree from the University of California at Berkeley. Dr. Shim has 40 books to his credit and has published over 50 articles in accounting and financial journals including *The CPA Journal, Advances in Accounting, International Accountant,* and *Financial Management.*

JOHN P. WALKER, Ph.D., CPA, is Professor of Accounting and Information Systems at Queens College and a computer consultant. He served as Chief Financial and Administrative Officer at the New York City Economic Development Corporation, was Executive Vice President at Data Spectrum, Inc., and served as assistant corporate controller at UNC, Inc. Dr. Walker also acted as a systems analyst and programmer at banks and was a management consultant to KPMG Peat Marwick, CPAs. His articles have appeared in such journals as the *Journal of Accountancy, National Public Accountant, Ohio CPA, Management Accounting, Virginia Accountant Quarterly,* and *Internal Auditor.*

ANIQUE AHMED QURESHI, Ph.D., CPA, CIA, is a consultant in accounting and computer sciences and professor of accounting and computer science at the University of Tampa. Dr. Qureshi has authored books for The Glenlake Publishing Company, Ltd., Prentice-Hall, American Management Association, and has contributed chapters to books for McGraw-Hill. His articles have appeared in *Accounting Technology, The CPA Journal, National Public Accountant, Management Accounting,* and *Internal Auditing.* He has made presentations at the American Accounting Association, American Association of Accounting and Finance, and INFORMS.

SUSANNE O'CALLAGHAN, Ph.D., CPA, CIA, is an Associate Professor of Accounting and Information Systems at Pace University. She is also an actuary and computer consultant. Dr. O'Callaghan was formerly a partner in Hubbard and O'Callaghan, CPAs and Director of Internal Auditing at the University of Nebraska. Her articles have appeared in many journals including the *Journal of Accountancy* and *Internal Auditor.*

PAUL KOKU, Ph.D., is Associate Professor of Accounting and Information Systems at Florida Atlantic University and a computer consultant. His articles have appeared in the *Journal of Business Research, International Journal of Technology Management, Journal of Market Focus Management,* and *Journal of Marketing.*

What This Book Will Do For You

The purpose of this book is to assist business people to understand artificial intelligence software and to make practical use of it. The topic of artificial intelligence (AI), including expert systems and neural networks, will keep business people up-to-date on the latest and most significant computer software applications. The book will be of immeasurable benefit to financial managers, accountants, tax managers, marketing managers, business managers, credit managers, loan officers, insurance executives, production managers, attorneys, economists, and others in the business world.

Artificial intelligence is the field of computer science that studies how machines can be made to act intelligently. It is the use of software to simulate the functions of a decision-maker's mind in carrying out his or her daily job responsibilities. The *Handbook* clearly defines and explains artificial intelligence and how it is used. It discusses the use of artificial intelligence by those directly or indirectly involved in business activities and shows how the use of AI can aid in making better and faster business decisions.

The business person must have an understanding of AI applications—expert systems, fuzzy logic, and neural networks—to properly conduct functions and maximize the company's growth potential. Artificial intelligence is the latest cutting edge technology. This knowledge will give business managers a competitive edge— potential mergers and acquisitions may be evaluated with expert systems and neural network software. Emerging trends in AI are discussed so the executive can properly plan for the future.

Difficult topic areas have been written in simple terms. The book is a practical "how to" guide—a what to look for, what to do, and how to apply what has been learned. Examples to illustrate practical applications as well as step-by-step instructions are provided. These include sample worksheets, schedules, checklists, charts, exhibits, tables, graphs, diagrams, forms, case studies, and computer printouts. Answers to commonly asked questions are given. The uses of this *Handbook* are as varied as the topic areas presented.

The glossary defines terminology used in AI. A particular area of interest may easily be found by referring to the index.

In financial accounting, artificial intelligence software assists in tracking and monitoring accounts, establishing reserves, and preparing and analyzing financial statements.

In managerial accounting, AI assists in cost management and analysis, planning and control, budgeting, forecasting, capital budgeting, resource allocation, variance analysis, cash flow appraisal, and analysis of products and services.

In audits, AI aids in performing audits, internal control evaluation, obtaining evidence, fraud investigation, sampling, and risk analysis (e.g., predicting a "going concern" problem).

In taxation, AI software aids in tax planning, estate planning, determining the tax consequences of proposed transactions, identifying tax problems, tax research, and ascertaining tax compliance to tax law and rules.

Artificial intelligence aids CPAs in performing personal financial planning services for clients— portfolio management, including stock and bond selection, diversification, establishing different portfolio designs under multiple constraints, thorough analysis of companies, investment timing, trading activities, and hedge and arbitrage strategies. AI software is also applicable to financial planning with regard to asset and debt management, insurance selection and policy limits, and banking arrangements.

Accountants may use artificial intelligence software when conducting management advisory services for clients—giving advice of how accounting information and reports may be of interest to and used by marketing managers, credit managers, production managers, insurance managers, and international business and finance managers making global trading decisions. In addition, how to incorporate artificial intelligence in other management advisory services to provide real "value added" services is discussed and demonstrated.

In practice management, AI assists in establishing client profiles, scheduling staff assignments, pricing of services, and computer-aided instruction.

The technologies used in AI include expert/knowledge systems, neural networks, case-based reasoning, pattern matching, machine learning, and fuzzy logic. Artificial intelligence software can be used to achieve voluminous business results, including:

- Lower costs to perform business activities.
- Increase in productivity and efficiency.
- Maximization of revenue.
- Maximization of resources.
- Optimal decision making and analysis.

An expert/knowledge system is a group of computer programs conducting a task at the level of a human expert. Expert systems usually have a huge database based on knowledge and experience gathered from experts in a particular area (e.g., financial accounting, managerial accounting, auditing, tax, personal financial planning). The expert system interacts with the user by repeatedly asking for relevant data until it is ready to derive a decision or conclusion. In other words, a sequential series of questions are asked designed to procure additional information in a reasoning process to help solve a business problem. Later questions are based on the answers to prior questions. The answers are not necessarily perfectly correct, but are logical conclusions. Besides reaching a decision, the expert system provides its underlying logic.

The major elements of an expert system are the database management system, the knowledge database, the inference engine, the domain database, the user interface, and the knowledge acquisition facility. Each of these components is fully discussed to achieve cost-effective optimal results.

A business application is a good candidate for developing an expert system if it involves expert knowledge, judgment, and experience. The business problem must be heuristic in nature as well as easily and clearly defined. Expert systems are appropriate for unstructured situations and tasks that are interactive.

The businessperson must be familiar with expert system shells that are a collection of software packages and tools used to design, develop, implement, and maintain expert systems. These are all discussed here.

Expert system development tools are available to simplify, facilitate, and establish, or enhance an expert system. The development aids include if-then rules, interfaces with databases, tools to better use spreadsheets and programming languages, and tools to generate the inference engine.

Areas of expert system usage include:

- Financial analysis.
- Preparing budgets and forecasts.
- Analysis of accounts.
- Planning capital and staff.
- Appraising loan applications.
- Security and control.
- Audit planning.
- Preparation and analysis of reports.
- Reduction of risk.
- Investment and portfolio management.
- Appraisal of internal control.
- Merger and acquisition analysis.
- Project scheduling.
- Product and service evaluation.
- Tax planning.
- Identify areas of fraud.
- Identify manufacturing deficiencies.

Internal accounting systems are ideal for expert system applications. Expert systems can be developed to appraise cash flows, accounts receivable, accounts payable, and management control. The knowledge base may include accounting, audit, and tax rules from the American Institute of CPAs, Financial Accounting Standards Board (FASB), Securities and Exchange Commission (SEC), and Internal Revenue Service (IRS).

In external (outside) auditing by the independent CPA, expert systems can apply accounting standards consistently when preparing accounts or conducting audits. Expert systems are very useful in auditing—selecting audit programs and/or test samples, determining error levels, conducting analytical reviews, and formulating a judgment based on the findings. Expert systems may also assist in internal control appraisal, evaluation of risk, disclosure compliance, scheduling audit assignments, and appraising auditor behavior. Expert systems can provide significant evidence that client reports are in conformity with Generally Accepted Accounting Principles (GAAP).

Taxation has a complex set of rules and procedures suitable for expert systems. Tax expert systems may be used for estate planning, tax research, determining the tax consequences for stock redemptions, and other complex tax issues associated with various transactions.

A loan analysis expert system can either accept or reject the application for loans and credit. Acceptance can be conditioned on some predetermined criteria. The expert system can identify questionable loans in terms of default risk.

With neural network (NN) technology, the computer learns from actual observations of characteristics and outcomes from databases and develops an ability to predict outcomes from the characteristics presented. The system can be designed to continue to learn when more observations become available. This building of observations allows the NN to make better decisions in the future. Neural networks simulate human intelligence and learn from experience.

NNs convert the computer into a "thinking" problem solver. A NN program can be used to confirm auditor judgments. There are four types of neural networks discussed in this book (1) prediction, (2) classification, (3) data filtering, and (4) optimization. Neural networks learn continuously and network relationships can be modified with changes in inputs and outputs coupled with retraining the network. Neural networks can create new knowledge in response to change in the environment. NNs are flexible because they are not restricted by a predetermined set of conditions.

NNs can be used to appraise portfolio management—they can scan for non-performing and overvalued stocks and predict stock patterns. The most common application of NNs is in the domain of pattern recognition. For these applications, NNs are useful in analytical procedures, internal control assessment, fraud risk assessment, and identification of clients in severe "financial distress."

Artificial intelligence systems can use either serial processing or parallel processing. A serial processing system makes only one decision at a time while a parallel processing system can make several concurrent decisions. Expert systems use serial processing and perform well in organized step by step processes. Neural networks use parallel processing systems and perform well with multiple and nonserial inputs.

An expert system processes a series of if-then rules, matching each "if" with the respective "then", until a final result is accomplished. This matching process is inherently sequential. NNs, on the other had, use parallel processing and are able to concurrently appraise multiple input.

Expert systems use deductive reasoning contrasted with neural networks that rely on inductive reasoning. Knowledge in the form of predefined rules is entered into the system and specific actions or conditions are derived from these rules. A neural network acquires its knowledge in the form of examples. General conclusions are derived from specific instances.

NNs offer some advantages over expert systems. For example, neural networks are self-adapting and can learn from information to uncover hidden patterns and relationships.

Fuzzy logic can be viewed as a language allowing for the translation of sophisticated statements from natural language into a mathematical formulation. Fuzzy logic relates output to input, and people do not need to comprehend all the intermediate steps. The inputs are provided by sensors rather than by people. Fuzzy logic requires fewer rules than expert systems and fewer data values than neural networks. It uses linguistic, not numerical, variables. Fuzzy logic simplifies knowledge acquisition and representation. It may be used to enable greater flexibility and more options. Fuzzy logic is beneficial in developing acceptable solutions to client problems with a wide range of potential actions and a high level of uncertainty.

Fuzzy logic is particularly useful in contract analysis and decision marking. In addition, business people use fuzzy logic in rating the safety of investment securities.

An intelligent agent is software that can conduct intelligent functions. The major attributes of intelligent agents are autonomy, communication, cooperation, reasoning, and adaptive behavior.

In conclusion, the *Handbook* shows clearly how artificial intelligence software can aid business people by enabling them to conduct their activities more efficiently and effectively. It provides all the information to successfully use expert systems and neural network software.

CHAPTER 1

Introduction

The aim of artificial intelligence (AI) is to create machines that can perform complex tasks as well as, or better than, humans. In order to perform those complex tasks, machines must be able to perceive, reason, learn, and communicate.

Business professionals need an understanding of artificial intelligence applications, especially the expert systems and neural networks that are being used extensively by large companies. The purpose of this book is to help business professionals understand AI concepts so they can make practical use of AI.

Most AI experts recognize that achieving intelligent machines remains a distant goal, even though in pursuit of that goal, useful machines and systems have been created and are being used every day. For example, AI systems are used routinely to evaluate creditworthiness of both companies and individuals. However, a truly intelligent machine capable of thinking, reasoning, and performing complex tasks may not exist for quite some time.

Artificial intelligence applies human reasoning techniques to computers. The software and hardware are designed to simulate the human mind. Expert systems and neural networks are just one application of AI. Expert systems are computer programs that reflect the behavioral characteristics of hired experts who are specialists in a given field. The software is designed to emulate how the experts would make decisions to solve clearly defined problems. An expert system provides advice using its knowledge and experience base. It is designed to ask for additional information in a reasoning process directed to solving a specific business problem, such as how to reduce a given cost (e.g., a manufacturing cost) while improving productivity and quality. It is also appropriate for unstructured situations and tasks, is interactive, and uses judgment.

CONVENTIONAL VERSUS ARTIFICIAL INTELLIGENCE APPLICATIONS

Conventional computer applications such as word processing and spreadsheets manipulate symbols and numbers step by step. Their flow tends to be linear: The actions that must be taken to solve a given problem are clearly defined.

In contrast, because knowledge-based expert systems are designed to emulate human reasoning, they tend to have many more branches and nonlinear approaches to solving problems. And because AI systems are more complex, they are harder to develop.

Knowledge-based expert systems are designed to capture information about relevant objects or events. Their structure may range from interactive to fully automated.

The outputs of both conventional and AI programs often look the same. Both use an algorithm of some sort and they are often coded using similar programming languages. Where the two types of programs differ is at the conceptual level.

An AI application is only as good as what it knows. The knowledge base is typically derived from human experts. A potential problem is that a human expert who may be very good at solving particular types of problems is not necessarily very good at defining the rules and reasoning she used to solve the problem.

EXPERT SYSTEM SHELLS AND PRODUCTS

A "shell" is a collection of software packages and tools used to design, build, implement, and maintain AI expert systems for a company. Shells may be generic (already prepared off-the-shelf for use in final form) or customized (requiring special preparation). In either case, the user enters

Introduction

the relevant information and parameters, and the expert system generates the solution to the problem.

Expert system development tools are also available to simplify and facilitate the establishment or enhancement of an expert system. These aids include if-then rules, interfaces with databases, tools to improve use of spreadsheets and programming languages, and tools to generate the interface engine.

The expert system asks the user a sequence of logical questions. Each question is based on the answer to the previous question. After all queries have been asked and answered, the expert system draws its conclusions.

BUSINESS APPLICATIONS OF ARTIFICIAL INTELLIGENCE

Expert systems are being used for:
- Accounting.
- Appraisal of internal controls.
- Auditing.
- Claims authorization and processing.
- Competitive analysis.
- Configuration and organizing.
- Credit authorization.
- Data communications.
- Determining the adequacy of expense provisions and revenue sources.
- Financial analyses.
- Manufacturing and capacity planning.
- Repair and maintenance reports.
- Report preparation and analysis.
- Resource planning.
- Risk evaluation.

- Routing.
- Scheduling.
- Security.
- Strategic marketing.
- Strategic planning.
- Transaction processing.

Accounting Applications

Expert systems can consistently apply standards for preparing accounts and conducting audits. Expert systems are ideal for:

- Appraising cash flows.
- Managing accounts receivable.
- Managing accounts payable.

Expert systems provide decision models for planning and control. They may also be used to schedule production and do inventory analysis. In accounting, expert systems can help to maintain ledgers, analyze revenue (by price, volume, or mix of product/service), prepare payrolls, analyze costs (e.g., by category and type), prepare and analyze financial statements, prepare working papers, do compliance reporting, prepare budgets and forecasts, age customer accounts, analyze merger and acquisition options, convert between accounting bases (e.g., accrual to cash), and decide whether to refinance debt.

In auditing, expert systems can reduce cost and time required to make audit decisions, improve the audit plan, and do substantive testing. Expert systems can select or create an audit program, conduct an analytical review, analyze data and evidence, select a sample and test data, determine an error rate, schedule and monitor the audit engagement, uncover illogical relationships between accounts, and evaluate assets (e.g., cash) and liabilities (e.g., accounts payable). Expert systems help users evaluate internal controls, appraise risk, determine compliance with disclosure requirements, analyze auditor behavior, determine whether the accounting and reporting system continue to meet the intended objectives, and select audit software and hardware for a particular task.

In managerial accounting, expert systems may be used in making capital budgeting decisions, such as selecting the right asset, the appropriate mix of products and services, and whether to keep or sell a business segment, or buy or lease equipment.

Multinational companies may use expert systems in assessing accounting, reporting, and legal requirements. The expert system gives the company advice on appropriate account types, reporting formats, and conditions in the global financial market that may affect the company's financial condition.

In practice, the firm may use expert systems to make decisions about staff development and assignment. In taxation, expert systems are used in tax planning and preparation, and to facilitate compliance with tax rules and procedures. Expert systems may also be used in compliance matters, as in determining if the company is in compliance with government regulatory or legal requirements so as to avoid penalties.

Finance Applications

Expert systems help in assessing risk (e.g., theft, fire, flood). An insurance company may use expert systems to evaluate and process claims, including highlighting suspicious claims for possible fraud. Expert systems can be optimally based on such factors as cost, time, availability, risk, and demand patterns.

In investment analysis, expert systems can recommend appropriate investments, taking into account such factors as the state of the economy, the user's risk preferences, tax rate, liquidity, dividend payout, capital appreciation (depreciation), desired portfolio mix, constraints and limitations, and SEC requirements. Expert systems make it possible to time buys and sells of securities because they consider multiple real-time data, both external and internal.

A "rule generator" expert system identifies information patterns and generates trading recommendations. A "critic" expert system analyzes system-recommended trades and explains the results. An expert system can provide 24-hour trading programs so as to optimize opportune conditions in both domestic and international markets, such as changes in foreign exchange rates.

Expert systems can recommend hedges to reduce a company's investment risk, such as futures contracts, options, and swaps. They can also appraise arbitrage opportunities and trigger transactions.

An expert system can make a decision whether to grant a loan and if so, how much it should be. It considers profitability, risk, economic conditions, and management policy. The expert system may approve the loan subject to certain criteria and restrictions. It can choose the appropriate interest rate, line of credit, repayment schedule, and collateral requirements. It can identify existing questionable loans.

Marketing Applications

Expert systems may be used in market planning and research, making strategic marketing decisions for both products and services, new product development and enhancement, warranty service planning, choosing product features and options, goal formulation (price, volume, profit), determining the marketing mix of products and services, return policy, advertising and promotion, deriving customer profiles, setting prices, establishing discount and credit terms, sales representation, deriving the best distribution channel system, product quality appraisal, and formulating the best style and packaging. The knowledge base of this type of expert system includes market structure, customer characteristics, and competition.

WHAT IS A MACHINE?

The word *machine* means different things to different people. A machine may be made of a material like silicone. It may be made of proteins. Some artificial intelligence experts believe that thinking can occur only in living machines made of protein.

Symbolic Processing

Other artificial intelligence experts argue that the physical substance of the machine does not matter. The essential ingredient is the ability to manipulate symbolic data. Computers can add numbers, sort data, perform logical comparisons, and replace one symbol with another.

Knowledge-based expert systems that require extensive knowledge of the domain have been built using the symbol processing approach. The knowledge the machine needs is specified using symbolic structures, on which operations are then performed. Symbol-processing approaches typically have a "top-down" design. They begin at the top, the knowledge level, proceed down towards the symbol and then the operations level.

Knowledge-based expert systems may interact with users in formulating optimal decisions. The expert system continually asks users for facts until it has enough to make the decision. It will also explain to the user its logic in making that decision. The expert system continues to learn more and more as it receives additional information and answers to its questions so it can make better decisions.

The six key components in an expert system are:

1. A knowledge database of rules, cases, and criteria for making decisions.

2. A domain database of information in the specific area of study.

3. A database management system to manage the other two databases.

4. An inference engine (processing system) comprising the interface strategies and controls experts use to manipulate the first two databases. This is the brain of the expert system. It receives the request from the user interface and conducts reasoning in the knowledge base. The inference engine aids in problem solving by processing and scheduling rules, among other activities. It asks for additional data from the user, makes assumptions about the information, and formulates conclusions and recommendations. The inference engine may also determine the extent to which a recommendation is qualified and rank multiple solutions. Its engine can aid in many key decisions, such as choosing financial accounting approaches and managing assets and debt.

5. The user interface. The explanatory features, on-line help, debugging tools, etc., that help the user understand and get the most advantage out of using the expert system.

6. A knowledge acquisition facility that enables interactive processing between the user and the system, and enables the system to obtain the knowledge and experience of the human expert.

Subsymbolic Processing

Some AI experts believe that the most important ingredient of AI intelligence is its ability to perform *subsymbolic,* not just symbolic, processing. The subsymbolic approach has a "bottom-up" design. Subsymbolic processing has the ability to process signals as well as symbols. For example, the human ability to recognize a face is hypothesized to depend on processing images as multidimensional signals, not just symbols. Neural networks are an example of the subsymbolic design approach to AI. Neural networks allow computers to learn from a database; a neural networks system obtains positive or negative responses to output from the operator and stores that data for later use in making better decisions.

Many neural network software programs on the market are ideal for business applications. Neural network software converts the order-taking computer into a "thinking" problem solver. This allows computers to make some of the mundane decisions previously made by accountants, such as deciding on the type of lease or the type of construction accounting method.

Neural network software, like human intelligence, learns from experience. For example, each time a neural network program makes the correct decision (predetermined by the human expert) in recognizing a sequential pattern of information, the programmer reinforces the program with a stored confirmation message. On the other hand, if the decision is incorrect, the confirmation message is negative. Thus, over time, experimental knowledge in a subject is built. The many applications of neural networks include:

- Analyzing customer behavior patterns.
- Assessing credit.
- Examining spending patterns.
- Internal auditing.
- Managing and analyzing investment portfolios.
- Managing working capital.
- Predicting bankruptcy.
- Scanning customer purchases.

Neural network technology has also been used to predict returns on bonds, interest rates on Treasury bonds, stock market movements, and currency exchange rates. A portfolio manager can use neural networks to identify nonperforming or undervalued securities. Many banks have successfully used these systems to control credit card fraud by recognizing fraud based on past charge patterns.

Neural networks can be applied in situations where traditional techniques have not yielded satisfactory results. They are also ideal where a small improvement in modeling performance can have a significant effect on operational efficiency or profits, as with direct marketing applications. The response rate in direct marketing is typically quite low. Using a neural network to analyze demographic data could improve the response rate by a few percentage points, significantly reducing costs.

GENETIC ALGORITHMS

Genetic algorithms are part of an increasingly popular AI area known as evolutionary computing. They are based on Darwin's theory of evolution—using genetic algorithms, a solution to a problem is "evolved."

An algorithm is started with a set of solutions, represented by chromosomes called "population." A new population is formed by taking solutions from an older population in the hope that the new population will be better than the old one. Solutions selected according to their "fitness" form the offspring. The better they are, the better their chances of reproducing. This process is repeated until some predefined condition is satisfied.

An example of a genetic algorithm program is *Generator,* a program that has several versions, including *Generator TFV* (The Financial Version), which is designed to interact with Microsoft Excel worksheets. Generator TFV may be used to solve complex problems that otherwise might require a tedious trial-and-error approach. Generator TFV can:

- Optimize stock or bond portfolios for return/risk tradeoffs.
- Structure portfolios to track specific indexes at the lowest cost.
- Select market-trading strategies.
- Identify efficient transactions for balancing a portfolio.
- Optimize the yield/return mix within portfolio constraints.
- Analyze alternative tax strategies.

For instance, a spreadsheet may be used to analyze stock or bond portfolios. "What-if" analysis helps the user decide whether to add and subtract securities or assets by calculating whether it is possible to increase returns or reduce risk. As the portfolio grows, so do the number of constraints. This makes it less likely that an optimal solution will be found using trial and error. Generator TFV[1] may be used to optimize any financial problem that can be set up on an Excel worksheet.

TURING'S TEST

Turing's test, which is named for Alan M. Turing, a pioneering researcher in artificial intelligence, may be used to determine if a machine is able to simulate human intelligence. (Appendix 1-A presents the results of a survey about how people perceive a machine's ability to act intelligently.) At present, no machine is capable of passing the Turing's test, and it is unlikely that one will pass the test in the near future. Turing [1950] described the test as follows:

> It is played with three people, a man (A), a woman (B), and an interrogator (C), who may be of either sex. The interrogator stays in a room apart from the other two [communicating with them via teletype]. The object of the game for the interrogator is to determine which of the other two is the man and which is the woman. He knows them by labels X and Y, and at the end of the game, he says either "X is A and Y is B" or "X is B and Y is A." The interrogator is allowed to put questions to A and B thus:
>
> C: Will X please tell me the length of his or her hair?
>
> Now suppose X is actually A, then A must answer. It is A's object in the game to try and cause C to make the wrong identification.
>
> The object of the game for the third player (B) is to help the interrogator.
>
> We now ask the question, "What will happen when a machine takes the part of A in this game?" Will the interrogator decide wrongly as often when the game is played like this as he does when the game is played between a man and a woman? These questions replace our original, "Can a machine think?"

---

[1]Generator TFV is available from New Light Industries, Ltd.

Introduction

The original Turing test involved a human interrogator using a computer interface (such as e-mail or a chat room on the Internet) to communicate with another human and a machine (see Figure 1-1). If the human interrogator is unable to determine which of the two sources is a machine, the machine is said to *simulate* human intelligence. An intelligent machine would be able to "trick" the human interrogator into believing that the machine is really a human. (See Appendix 1-B for a sample of Turing's test questions.)

Figure 1-1. Turing's Test

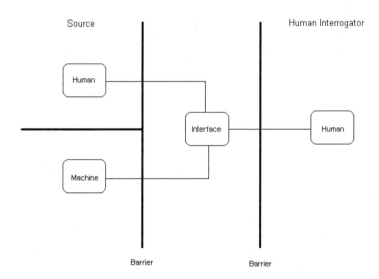

The Turing test does not allow the interrogator to physically observe whether the source is a machine or human. The interrogator is allowed to observe only the intellectual behavior through communication.

Turing proposed that about five minutes should be sufficient to make the determination, but the exact amount of time is generally considered irrelevant. It is possible that the human interrogator can randomly guess which source is a machine and which is human. Therefore, we expect that if the machine is truly intelligent, the human interrogator's accuracy rate will be 50 percent. This result should hold no matter what people are involved in the test.

ELIZA

Eliza is a computer program created in the mid-1960s that displays some artificial intelligence. Eliza can appear to engage a human in conversation, without really comprehending what the human is saying. Eliza asks the human user a question, and uses the answer received to formulate yet another question. The program stores subject information in its data banks and uses it to converse about a variety of topics. Eliza is also able to detect speech patterns. The following passage shows Eliza talking to a teenage girl. The teenager speaks first, and Eliza's answers are in CAPITALS.

Men are all alike.

IN WHAT WAY

They're always bugging us about something or other.

CAN YOU THINK OF A SPECIFIC EXAMPLE

Well, my boyfriend made me come here.

YOUR BOYFRIEND MADE YOU COME HERE

He says that I am depressed all the time.

I AM SORRY TO HEAR YOU ARE DEPRESSED

It's true that I am unhappy.

DO YOU THINK COMING HERE WILL HELP YOU NOT TO BE UNHAPPY

I need help, that seems certain.

WHAT WOULD IT MEAN IF YOU GOT SOME HELP

Perhaps I could learn to get along with my mother.

TELL ME MORE ABOUT YOUR FAMILY

My mother takes care of me.

WHO ELSE IN YOUR FAMILY TAKES CARE OF YOU

My father.

YOUR FATHER

You are like my father in some ways.

WHAT RESEMBLANCE DO YOU SEE

You are not very aggressive but I think you don't want me to notice that.

WHAT MAKES YOU THINK THAT I AM NOT VERY AGGRESSIVE

You don't argue with me.

WHY DON'T YOU THINK I ARGUE WITH YOU

You are afraid of me.

DOES IT PLEASE YOU TO BELIEVE I AM AFRAID OF YOU

My father is afraid of everybody.

WHAT ELSE COMES TO MIND WHEN YOU THINK OF YOUR FATHER

Bullies.

DOES THAT HAVE ANYTHING TO DO WITH THE FACT THAT YOUR BOYFRIEND MADE YOU COME HERE

The conversation with Eliza gives the appearance that Eliza is understanding the teenage girl. In fact, Eliza is communicating by simply modifying the human user's sentence and using it as a response. To experience communicating directly with Eliza, visit *http://www-ai.ijs.si/eliza/eliza.html.*

PROCESSING

AI systems can use either serial processing or parallel processing. While a serial processing system makes only one decision at a time, parallel-processing systems can make several concurrent decisions. Expert systems use serial processing systems; neural networks use parallel processing systems.

An expert system processes a series of if-then rules, matching each "if" with the respective "then" until a final result is achieved. This matching process is inherently sequential. Because neural networks, in contrast, use parallel processing, they are able to concurrently evaluate multiple inputs.

BRANCHES OF AI

There are many branches of artificial intelligence, including:

- *Automatic Programming.* Once the capabilities of a computer program are described, the AI system writes the code for it.
- *Bayesian Networks.* The use of probabilistic information by an AI system in making inferences.
- *Machine Learning.* Computer programs that learn from experience.
- *Natural Language Processing.* Processing and possibly understanding the natural human language.
- *Speech Recognition.* Converting oral speech into written text.
- *Visual Pattern Recognition.* Responding to visual stimuli in a way similar to the human sense of sight.

PROGRAMMING FOR AI

While AI programs have been written in almost every programming language, the most popular are Lisp, Prolog, C/C++, and Java.

- Lisp, a high-level language, offers the advantage of fast prototyping—at the expense of slower execution. Universities and laboratories use Lisp extensively for AI research. Some important features of Lisp are:

  –Dynamic typing
  –Extensibility
  –Functions as data
  –Garbage collection
  –Interactive environment
  –Uniform syntax

- Prolog, another high-level language, is very popular for AI programming especially where logic plays a significant role. Prolog is difficult to learn.
- C/C++, a general-purpose language, is used extensively for both AI and non-AI applications. C/C++, which is very fast, is often used when speed is of paramount importance. For instance, back-

propagation in neural networks, which requires fast execution, can be handled simply in C/C++.

- Java syntax is like that of C/C++. Java should be used when portability among different platforms is of prime importance. Java also offers automatic garbage collection like Lisp. However, code written in Java does not execute as fast as code written in C/C++.

BACKWARD CHAINING

Many expert systems rely on the backward chaining procedure. Backward chaining uses rules and deductive reasoning to logically draw conclusions. The process starts by picking the most likely hypothesis and works backward to prove it. Backward chaining relies on rules such as "If A is true, then B must be true." In the form of an if-then statement, the if part is called the "premise" and the then part is called the "conclusion." The collection of if-then rules form the system's knowledge base.

Physicians making a diagnosis often use backward chain reasoning. They start off with the most likely diagnosis and perform tests to confirm it. Initial testing may do that or it may not, which means further testing is necessary. Sometimes, the initial diagnosis proves incorrect so the physician must form another diagnosis. The process continues until the physician is able to confirm a diagnosis.

Rule priority is an important part of backward chaining. The expert system must decide which rule to pick to prove the initial hypothesis.

BOTS AND AGENTS

The word *bot* is short for robot. Bots are software tools used to dig through data. Give a bot directions and it responds by bringing back the answer. An *agent* is a bot that goes out on a specific mission.

Bot software is ideal for performing methodical searches on the web. Web search engines routinely send out software robots that crawl from one server to another to compile the data the search engines need. Intelligent bots can learn from past experience and make decisions as they search data.

FUZZY LOGIC

Lotfi Zadeh of the University of California, Berkeley, introduced fuzzy logic. Fuzzy logic is used to model the uncertainty of natural language. It recognizes the concept of partial truth. That is, there exists something between "completely true" and "completely false." For example, consider the concept of "tallness." *Tall* is described by Zadeh as a *linguistic variable*. A fuzzy subset of TALL can be defined by asking the following question: *To what degree is a person tall?*

A degree of membership may be assigned to the fuzzy subset TALL using a membership function based on the person's height. Consider the following example:

tall(x)   =   { 0, if height(x) < 5 ft.,
(height(x) - 5 ft.)/2 ft., if 5 ft. <= height (x) <= 7 ft.,
            1, if height(x) > 7 ft. }

A graph of this function is as follows:

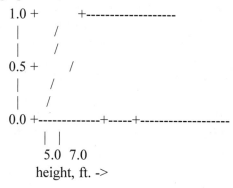

height, ft. ->

Given the membership function, here are some example values:

Person	Height	Degree of Tallness
Juan	3'2"	0.00 [I think]
Yoke	5'5"	0.21
Drew	5'9"	0.38
Erik	5'10"	0.42
Mark	6'1"	0.54
Kareem	7'2"	1.00 [depends on who you ask]

Expressions such as "Drew is TALL" = can be interpreted as degrees of truth. In this case the expression = 0.38.

THE JAVA EXPERT SYSTEM SHELL

The Java Expert System Shell (Jess)[2] is a rule engine and scripting environment written in Java. With Jess, it is possible to build intelligent Java applets and applications that can reason using knowledge supplied in the form of declarative rules. An algorithm called *Rete* is used to match the rules to the facts. Rete does not rely on simple if ... then statements in a loop. This improves the processing speed of Jess. Jess is available free for academic use. Commercial users need to buy a license.

INSTANT TEA

Instant Tea is both an integrated development environment (IDE) application and a Java Web program (delivery applet) for the production and delivery of fuzzy knowledge bases on the Web. Instant Tea makes it possible to build web-based knowledge bases without any programming.

A consultation with the system requires the user to answer questions the system generates. The delivery applet applies knowledge bases created by the IDE. The system uses the user's responses and the knowledge bases logical inferences to derive solutions to the problem and generate recommendations. The IDE and delivery features include:

- An object-oriented editing environment for creating classes, subclasses, attributes, and values.
- Support for customized interfaces for knowledge bases.
- Support for integration of a customized Java class to extend knowledge-based processing.
- An integrated rule editor and WYSIWYG (what you see is what you get) testing environment using Java's appletviewer.

---

[2] *See: http://herzberg.ca.sandia.gov/jess/* for more information on Jess or to download a copy of Jess and its user manual.

- Automatic propagation of changes to attributes and values to update rules.
- Rule creation by double-clicking attribute values.
- Support for most logical operations—"AND," "OR," "NOT," and "ELSE."
- Support for fuzzy numbers, fuzzy variables, and fuzzy reasoning.
- An automatic fuzzy membership function generator that supports hedges.
- Multiple comparison operators for numeric attributes (greater/less than, between/not between, etc.).
- Triangular and trapezoidal membership functions for fuzzy numbers.
- Rule groups to simplify maintenance.
- Support for prioritizing goals and rules.
- Support for confidence factors for individual rule premises and conclusions.
- Ability to set inference strategy to pursue single or multiple values for goals and to set confidence thresholds for firing rules.
- Support for customized user questions, explanations, and conclusions in text or HyperText Markup Language (HTML).
- Easy and efficient association of graphics with questions or an individual attribute's value.
- Automatic output of text-based knowledge bases that is easy to edit and spell-check with your favorite word processor.
- Automatic generation of HTML files for delivering customized applets on your web.
- Option to create customized delivery interfaces by specifying layouts, text size, colors, and applet properties.
- Open connections to Java.
- Ability to connect knowledge bases to web databases.

- An intelligent queries function.
- The option to add the functionality of Java technologies to knowledge bases.
- Delivery of customized traveling expert advice from a web site.
- Easy set-up on a web server.
- Uploading of Instant TEA's class files, knowledge bases, and graphics.
- Indication of directory locations by setting a few applet parameters.
- An intuitive point and click user interface.
- Automatic detection and presentation of HTML-based explanations and questions.
- Recommendations of web-based sites that users can automatically access for further clarification.
- Ability to trace rules and user responses to help explain how the system derived its conclusion.
- A platform-independent JAVA applet.
- Full integration with TEA's IDE and e-mail installation support for registered sites.
- Three-tier applications with connectivity to databases.

REFERENCES

Cox, E., *The Fuzzy Systems Handbook,* Academic Press, 1999.
Fayyad, U., *Advances in Knowledge Discovery and Data Mining,* Cambridge, MA: MIT Press, 1996.
Rogers, J., *Object-Oriented Neural Networks in C++,* Academic Press, 1997.
Turing, A., "Computing Machinery and Intelligence," *Mind,* 1950.
Winston, P., *Artificial Intelligence,* MA: Addison-Wesley, 1992.

APPENDIX 1-A: CAN COMPUTERS THINK?[3]

Alan Turing devised a test to determine if a computer could "think." He said, essentially, that if a person asked a question of a computer and receives a response that he could not differentiate from a human response, that computer could "think."

The following are the results of a survey:

1. About when do you believe that computers will be able to think?

When	Percent
Now	17.0
2000	02.0
2005	07.5
2010	19.0
2015	11.0
2020	04.0
Later	21.0

2. Do you believe computers today can think under Turing's definition?

Yes	28.3%
No	71.7%

3. Do you believe computers will ever be able to think?

Yes	71.7%
No	23.8%

4. Would you be comfortable having a thinking computer work for you?

Yes	77.4%
No	22.6%

5. Would you be comfortable working for a thinking computer?

Yes	34.0%
No	66.0%

---

[3]This appendix was taken from *http://users.erols.com/msagi/delphi-ai.html*. The survey was part of research on emerging technologies at The George Washington University, Washington, D.C.

6. Your education is

   Mean: 17.2 years of formal education

7. Your age is

Group	Percent
<20	34.8
21-30	30.4
31-40	13.0
41-50	17.4
51-60	02.2
61-70	02.2

8. Your gender is

Male	76.1%
Female	23.9%

A Pearson correlation of 0.500 at the 0.05 level (two-tailed) was found between years of respondent formal education and estimation of when computers can be expected to "think."

APPENDIX 1-B: QUESTIONS FOR TURING TEST[4]

The following is a listing of sample questions that have been proposed to discriminate between a machine and a human:

- Describe something that makes you anxious or nervous and explain why it does?
- Do you have a name?
- Have you ever been in love?
- How old are you?
- How tall are you?
- How would you feel if you saw a large man beating a small kitten?
- Is there anyone you distrust?
- Today is December 31, 2002. What is tomorrow's date? And why are you laughing?
- What was your most embarrassing moment?
- What's the difference between human beings and computers?

---

[4] This appendix contains a sample of Turing's test questions from *http://www.badpen.com/cgi-bin/turing/question.cgi*. See the web site for the full list.

CHAPTER 2

Artificial Intelligence and Expert Systems

Commercial application of AI technology has grown rapidly in the past decade. The technologies used in AI applications include expert/knowledge systems, neural networks, case-based reasoning, pattern matching, machine learning, and fuzzy logic.

Among all different AI applications, expert systems (ESs) are the most promising and have received the most attention from the commercial world. The problem domains ESs cover range from accounting, finance, management, and marketing to law and engineering. ES AI applications include problem-solving, planning, search, interpretation, training, monitoring, and control, to mention only a few.

Expert systems are computer programs that exhibit the behavior characteristics of experts; their computer software emulates the way managers and employees solve problems. Like a human expert, an expert system gives advice by drawing upon its own store of knowledge and by asking for information specific to the problem at hand. ESs are not exactly the same thing as decision support systems (DSSs). A DSS is computer-based software that assists decision makers by providing data and models. It primarily performs semistructured tasks. An ES is more appropriate for unstructured tasks. Both DSSs and ESs can be interactive, but because of the way DSSs process information, they typically cannot be used for unstructured decisions that use nonquantitative data. Unlike expert systems, DSSs do not make decisions; they merely attempt to enhance decisions made by operators by providing indirect support without automating the whole decision process.

Some general characteristics indicate whether a given business application is likely to be a good candidate for an expert system. The application must require the use of expert knowledge, judgment, and experience. The business problem must be heuristic and must be defined clearly. The area of expertise required for the application must also be well defined and recognized professionally, and the organization building the ES must be able to recruit an expert willing to cooperate with the development team. The size and complexity of the application must also be manageable given organizational resources, available technical skills, and management support.

EXPERT SYSTEMS IN DETAIL

An expert system, sometimes called a knowledge system, is a set of computer programs that perform a task at the level of a human expert. ESs are created on the basis of knowledge collected on specific topics from human experts, and they imitate the reasoning process of a human being.

Management and nonmanagement personnel use ESs to solve specific problems, such as how to reduce production costs, improve worker productivity, or reduce environmental impact. The computer program, methodically using a narrowly defined domain of knowledge built into it, comes up with a solution problem much as an expert would.

The key is that the domain must be narrowly defined. No ES can at this point give useful answers about all questions—each is limited, as a human expert is limited, to a particular field. For example, one ES could not tell the controller both whether to lease or buy a piece of equipment based on the tax differences and also how a pending business combination should be accounted for.

Among the many reasons why companies are building expert systems, a major one is to preserve expertise. Expert systems store tremendous amount of experts' knowledge, which are to be made available to people with less experience. Other reasons include the following:

- Improve productivity.
- Make expertise portable.

- Obtain otherwise unavailable expert advice
- Enhance the public image of the company

HOW EXPERT SYSTEMS WORK

Expert systems usually have the following six major components.

1. *Knowledge Database.* This database contains the rules and the cases used in making decisions.
2. *Domain Database.* This database contains the information relevant to the domain (the area of interest).
3. *Database Management System.* This system controls input and management of both the knowledge and domain databases.
4. *Inference Engine.* This component contains the inference strategies and controls used by experts to manipulate the knowledge and domain databases. It receives the request from the user interface and conducts reasoning in the knowledge base. It is the brain of the expert system.
5. *User Interface.* The user interface includes the explanatory features, on-line help facilities, debugging tools, modifications systems, and other tools to help the user use the system effectively.
6. *Knowledge Acquisition Facility.* This facility determines how the system acquires knowledge from human experts in the form of rules and facts. It allows for interactive processing between the system and the user. More advanced technology allows intelligent software to "learn" knowledge from different problem domains. The knowledge learned by computer software is more accurate and reliable than that of human experts.

The relationships of these components are illustrated in Figure 2-1. Based on these relationships, it is apparent that ESs must work interactively with system users to help them make better decisions. The system interacts with the user by continuously asking for information until it is ready to make a decision. Once the system has sufficient information, it returns an answer or result to the user. Note that not only must the system help make the decision; it must also give the user the logic it used to reach its decision.

FIGURE 2-1. Expert System Relationships

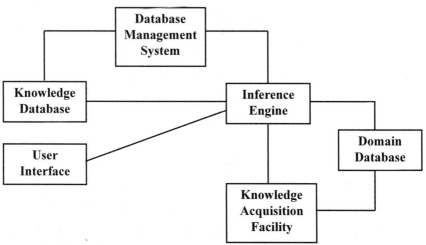

Thus, the *inference engine* processes the data the user provides to find matches with the *knowledge base* where the expert's information is stored. The *user interface* allows the user to communicate with the program. The *inference engine* (explanation facility) shows the user how each decision was derived. Expert systems are only as good as their programming. If the information in the knowledge base is incorrect or if the inference engine is not designed properly, the results will be useless. GIGO holds true—garbage in, garbage out.

ES TOOLS

Constructing viable expert systems requires powerful tools. The many ES tools on the market today vary widely in their functionality and hardware support requirements. One might start with a general-purpose AI programming language, such as *Lisp* or *Prolog,* to design and program parts of an ES from scratch. On the other end of the spectrum, it's possible to work with a large hybrid development environment that provides an ES shell, user interface builder, and other helpful facilities. The rest of this section will review and compare some popular commercial tools.

Languages for Expert System Development

The most basic ES development tool is a general purpose programming language. Lisp is the language most widely used. It has many features that ease the task of building any symbolic processing system. Moreover, Lisp is becoming more popular for conventional programming, particularly with the advent of Common Lisp and Lisp development environments that provide graphic user interface (GUI) support.

Prolog is another symbolic general-purpose programming language. Prolog is gaining in popularity, although it is far behind Lisp as an ES development language.

Conventional programming languages, such as C and C++, have also been used in ES development. Because the C language is so efficient and popular, many developers use Lisp for the prototype system and C to construct a final delivery system.

Using general purpose programming languages gives ES developers more flexibility in adapting the system to the problem domain. On the other hand, such languages are more difficult to apply because they give little or no guidance on how knowledge should be represented or how mechanisms for accessing the knowledge base should be designed.

Expert System Shells and Products

An alternative is to use an ES-specific tool, often referred to as an ES shell. An ES shell contains all the essential elements of an expert system except the domain-specific knowledge. The knowledge representation method and inference engine are built-in features that can be applied in constructing an ES for similar domains. To use an ES shell successfully, the domain characteristics must match those the shell's internal model expects.

There are a number of off-the-shelf ES shells that are complete and ready to run. The user enters the appropriate data or parameters, and the ES provides output to the problem or situation. Popular ES shells are described in Table 2-1.

A number of other tools make ES development easier and faster. These products help capture if-then rules for the rule base, assist with tools such as spreadsheets and programming languages, interface with traditional database packages, and generate the inference engine. With these, people with virtually no computer experience can run an expert system.

TABLE 2-1. Popular ES Shells

- *1st-Class Fusion* offers a direct easy-to-use link to the knowledge base. It also offers a visual rule tree, which shows graphically how rules are related.

- *Financial Advisor* analyzes capital investments in fixed assets, such as equipment and facilities.

- *Knowledgepro* is a high-level language that combines ES functions and hypertext. It allows for setup of classic if-then rules and can read database and spreadsheet files.

- *Leonardo* uses an object-oriented language called COMSTRAT, which marketing managers can use to help analyze the position of their companies and products relative to their competition.

- *Personal Consultant (PC) Easy* is used to route vehicles in warehouses and manufacturing plants.

APPLICATIONS OF EXPERT SYSTEMS

The use of expert systems is on the rise. Sales of expert system shells are increasing about 20 percent per year, with about 60 percent of the sales for use on IBM PCs or compatibles.

One of the main challenges to ES use is integrating ES concepts and functions into existing applications, including transaction processing. The business applications of expert systems are many and varied, including security, capacity planning for information systems, capital budgeting, financial analysis, product selection and mix, legal decision-making, management decision-making, and financial planning.

Some expert systems have been in existence for years. Among them are:

- *CoverStory* extracts marketing information from a database and automatically writes marketing reports.

- *Intelligent Scheduling and Information System (ISIS-11)* is used by Westinghouse Electric for scheduling complex factory orders.

- *CARGEX-Cargo Expert System* is used by Lufthansa to help determine the best shipping routes.

- The *NCR Corporation ES* for communications encodes an expert's knowledge into a form that can be used by a personal computer for more efficient analysis of difficulties regarding data communications.

- *ACE* is an expert system used by AT&T to analyze maintenance of telephone networks.
- *DELTA* is a General Electric ES that assists in engine repair.
- *XCON (Expert VAX System Configuration),* an ES with thousands of rules, was developed by Digital Equipment Corporation (DEC) to help configure and organize minicomputer systems.
- *Authorizer's Assistant (AA)* is an American Express ES used for credit authorization, weeding out bad credit risks, and reducing losses.
- *Watchdog Investment Monitoring System* is used by Washington Square Advisors, the investment management subsidiary of Northwestern National Life Insurance Company, to analyze corporate bonds to enhance clients' revenue. The analysis includes a change in financial ratios as an indicator of past performance and predictor of future financial directions.
- *Escape* is used by Ford Motor Company for claims authorization and processing.
- *Plan Power,* developed by Applied Expert Systems, analyzes a company's financial situation, then matches needs with the most appropriate financial products and services. The system will run scenario spreadsheets showing income tax ramifications, cash flows, net worth, and other critical factors based on alternative decisions.
- *GURU,* from Micro Data Base Systems, provides management advice and financial analysis using spreadsheets.
- A *Peat Marwick ES* is used to bring more consistency and precision to the auditing of commercial bank loans and provisions for bad debts.
- *XSEL,* DEC's sales support ES, reduces a three-hour system configuration/alternative generation task to 15 minutes and reduces the number of non-manufacturable systems specified from 30 percent to 1 percent. DEC claims it's worth $70 million a year.
- Canon's *Optex* camera lens design ES has made scarce, highly skilled lens designers 12 times more productive.

- The British National Health Services' *PC-Based Performance Analyst System* reduced the time required for an evaluation task from two hours to nine minutes—a factor of 80 in productivity gain.

Functional Area Applications

Expert systems are also used in accounting-related systems, capital resource planning, loan applications, strategic marketing, and developing strategic objectives.

Accounting Systems

Internal accounting systems are an ideal area for ES applications. The knowledge base can include information from accounting organizations like the American Institute of Certified Public Accountants (AICPA)—current tax laws, Securities and Exchange Commission (SEC) requirements, and Generally Accepted Accounting Practices (GAAPs). The inference engine for an accounting ES can help in many important decisions, including financial accounting approaches and management of cash flows.

Four areas of accounting in which ESs are particularly useful are (1) accounting standards, (2) taxation, (3) management and control, and (4) auditing.

An accounting standards ES would apply standards consistently in preparing accounts or performing audits. External auditors rather than internal auditors would probably perform this task more often.

The taxation area is restricted by complex rules and procedures. ESs make complying with these rules much easier, because all rules can be programmed into the computer. As we will explain later, tax planning has also benefited from ESs.

Management and control ESs supplement management information systems. Expert systems provide decision models for planning and control. As with any new management system, the internal auditor should evaluate potential benefits and control areas. The auditor must also periodically evaluate established ESs to determine whether they continue to meet the objectives they were designed to meet.

For auditing, expert systems can choose an audit program and a test sample, determine the level of error, perform an analytical review, and then make a judgment based on the findings.

Artificial Intelligence and Expert Systems

Expert systems can also be used to:
- Prepare working papers.
- Maintain ledgers.
- Prepare financial statements.
- Plan budgets and forecasts.
- Prepare and analyze payroll.
- Analyze revenues by volume, price, and product-service mix.
- Analyze expenses.
- Specify costs in terms of volume, price, and category.
- Convert accounts from cash to accrual basis.
- Age accounts receivable.
- Analyze financial statements.
- Review all financial aspects of the business.

Currently, there are only a few tax expert systems available, primarily for two reasons. First, if the information in the knowledge base is incorrect and bad decisions are made based on the system, the developer could be sued. Second, many ESs are built by large firms that want to protect their investment. It is not difficult to create an expert system using a shell. Other reasons may be that the tax code is constantly being revised (more change implies higher ES maintenance costs), some tax practitioners do not believe in the benefits of ESs, CD ROM tax databases already provide a lot of information, and tax-only expert systems are not sufficient alone to do business planning—more support is needed for planning decisions.

Among the few ES tax programs are *ExpeTAX,* a Coopers and Lybrand product used in tax planning and tax accrual. It uses a question-and-answer format to run a maze of 3,000 rules before outlining a client's best tax options. It will, for example, identify the differences between book and tax values. *Taxadvisor* is used for estate planning. *Corptax* examines the tax consequences of stock redemptions. Commerce

Clearing House (CCH) Inc. offers the *CCH Tax Assistant*. (While CCH does not call the product an expert system, it performs like one in many ways.) Tax Assistant uses user-entered information to make decisions. The software can reduce the time accountants spend calculating and generating reports and allows lower-level accountants to complete more difficult research tasks.

Some believe that expert systems are the future in tax accounting. ESs could be used for compliance work, for example, to determine whether an activity is passive. They could also be used to identify problems and make plans. For example, to determine whether a company is a personal holding company and how to avoid the associated penalty.

Capital Expenditures Planning

There are many capital investment decisions the company may have to make in order to grow. Examples are selection of the product line, whether to keep or sell a business segment, or lease or buy equipment, and which asset to invest in. Resource commitments may also be evaluated in such forms as new product development, market research, introduction of a computer, or re-funding of long-term debt. ESs may also be used to analyze mergers and acquisitions, whether to buy another company, or to add a new product line. *CashValue* is a commercially available capital projects planning ES.

Analysis of Credit and Loan Applications

A major function of any lending institution is making sound, profitable loans to business. Making too many risky loans can bring the institution large financial losses and perhaps even bankruptcy. Reliable loans to companies unlikely to default can substantially increase a bank's profitability. Because of the high degree of analytical skills and experience required, analysis of loan applications is a natural task for a computerized expert system. In extending loans and lines of credit to businesses the lender must take into account several key considerations, one of which is *management attitudes and style*. Will the management be able to grow in adverse as well as good times? How will management use the proceeds of the loan? Are there any potential problems with the company or management?

A loan analysis ES can either accept or reject applications for loans and credit. The acceptance can also be conditioned upon certain criteria. For example, the loan may be made only if the company receiving the funds agrees to make certain changes in its operation, management style, or marketing strategy. The ES can also identify default risk that makes a loan questionable. The result could be to make the loan at a higher interest rate or lower loan amount, with an altered repayment structure, or with higher collateral requirements. The decision might also be to reject the application.

GMAC has invested millions of dollars in its *Analyst* system, which evaluates the creditworthiness of General Motor's 10,000 domestic dealerships. The system is deployed in over 300 networked sites across the U.S. and it's estimated to pay back over $2 million a year.

Marketing Applications

Marketing ESs can allow marketing managers to make strategic decisions and plan activities. Establishing sales and profit goals, choosing which products and services to push, and profiling prospective customers are examples of marketing decisions that can be facilitated by ESs.

A marketing ES requires a knowledge base filled with data on customers, the structure of the market as a whole, diverse internal and external factors, and the competition. Once the general strategic marketing plan has been mapped out, the ES can explore specific goals, analyzing product quality and style, customer services offered, and any additional features and options.

Pricing policy decisions are equally important. The price analysis determines list price, discounts, and credit terms. Advertising, the role of sales representatives, direct marketing, publicity, the use of marketing research firms, and using professional marketing companies are also important decisions the ES can help make for promotional efforts. Finally, the system examines distribution channels for delivering products and services to customers.

Strategic Planning

Forecasting is a critical activity for most organizations. It is often costly and complex for reasons that include the multiplicity of forecasting methods and combinations, the absence of a forecasting method recognized to be the best, and the context dependence of methods based on available models, data characteristics, and the environment. In recent years AI-based techniques have been created to support a variety of operations management activities. One AI technique, rule induction, can be used to improve the accuracy of forecasts. Specifically, a rule-induction-based ES is trained with a set of time series data (the training set). Inputs to the ES include selected time series features and, for each time series, the most accurate forecasting method from those available. The ES is then used to recommend the most accurate forecasting method for a new set of time series (the testing set).

Applications in Finance

AI and ESs have many applications in banking, portfolio management, and trading of securities, among other areas of finance.

Banking

Banks have become big-time converts because ESs and AI are saving them a bundle. Among the member institutions of MasterCard International Inc., for example, ES programs designed to nip credit card fraud in the bud have prevented the loss of an estimated $50 million over the past 18 months. Other companies joining the bandwagon are Citibank and Deere & Co.

Insurance

ESs are operating throughout the insurance industry in the following areas, to name a few:

- *Underwriting.* Expert systems increase the consistency with which company standards are applied in evaluating risks like fire, flood, or theft.
- *Claims processing.* Fraud detection is particularly difficult in medical insurance because the claims are so complex. Expert systems substantially reduce labor cost by quickly evaluating claims and improving the detection of suspicious information.
- *Reserving.* Deciding how much reserves to set aside for future claims and ongoing payouts is like setting factory inventory sched-

ules—in both cases, and in many ways, ESs make it possible to consistently allocate resources to meet uncertain demands.

Portfolio Management

ESs are similarly useful in a variety of ways for portfolio management, as follows:

- *Security selection.* With over 100,000 stocks and bonds to choose from, choosing the right securities is a substantial challenge. There is far more information than can be intelligently digested. An ES analyzes data and makes recommendations. For example, Unitek Technologies' *Expert Strategist* analyzes financial statements.

- *Consistent application of constraints.* Managers of multiple portfolios must consistently apply a variety of constraints on different portfolios. The constraints include compliance with SEC rules, client guidelines, and consistency across related accounts. ESs helps financial professionals apply different portfolio designs under multiple constraints.

- *Hedge advisor.* The number of financial instruments, their complexity, and the complexity of their relationships are increasing rapidly. These instruments vary in margin, liquidity, and price. They can also be combined to create "synthetic securities" to hedge against market risk. An ES can help create these "synthetic securities."

Trading Advice

ES can be particularly helpful when the user is making decisions about which stocks to buy or sell. Some of these include:

- *Real-time data feed.* Timely information is critical in any trading application. An ES will integrate multiple real-time external and internal data sources into timely reports.

- *Trading rules and rule generators.* The conventional ES knowledge engineering approach to rule generators can replace writing. A rule generator ES will recognize data patterns and generate immediate hypotheses (trading rules) that lead to trading recommendations.

- *Critics and neural nets.* A critic ES will evaluate system-recommended trades, process explanations of those trades, and seek out culprits and heroes.

ESs in the Global Financial Market

There are a number of ways ESs can increase the likelihood of success for someone operating internationally. Such as:

- *Financial statement advice for multinational companies.* Multinational firms have unique reporting and legal problems. They must deal with such inconsistencies as varying reporting formats, regulatory requirements, and account types. An ES can solve these problems.

- *Twenty-four-hour trading programs.* Individual human traders cannot work for 24 hours—ESs can. With traderless expert systems, smaller companies have a better chance of entering global markets and trading on foreign exchanges.

- *Hedges.* In international security markets, many different types of hedges are possible, including interest-rate swaps, currency swaps, options, and futures. An ES can take advantage of those hedges.

- *Arbitrage.* ESs can quickly identify and evaluate arbitrage opportunities and trigger transactions.

BENEFITS AND DISADVANTAGES OF EXPERT SYSTEMS

Expert systems offer the following benefits:

- They increase output and productivity.
- They offer better accuracy, quality, and reliability.
- They can function as tutors, because they distill expertise into clearly defined rules.
- They capture scarce expertise.
- They share knowledge. An ES will be available to provide both second opinions within the domain and what-if analysis where results are sought on variable changes.

- Their decision time is shorter. An ES makes routine decisions rapidly.
- They enhance problem-solving capabilities.
- ESs are more secure than an expert employee, who may always be hired by a competitor.
- They reduce errors.
- They require fewer personnel.
- They require less training time for personnel.
- They improve decisions.
- They retain volatile or portable knowledge.
- They improve customer service.

But ESs also have drawbacks:
- They fail to adapt to a continually changing environment.
- They are usually confined to a very narrow domain and may have difficulty coping with broad decisions.

OTHER AI APPLICATIONS

Besides expert systems, major AI applications are fuzzy logic, automatic programming, natural language processing, and intelligent agents.

Fuzzy Logic

Fuzzy logic deals with uncertainty. This technique, which uses the mathematical theory of fuzzy sets, simulates the process of normal human reasoning by allowing the computer to behave less precisely and logically than conventional computers do. Fuzzy logic mathematics deals with nonprecise values within certain degrees of uncertainty. This technology allows logic to be used without precise information.

Applications of fuzzy logic can be found in controllers of appliances, among them rice cookers, VCRs, air conditioners, and cameras. In these products, fuzzy logic is used to make continuous small adjustments instead of merely switching a feature on or off.

Fuzzy logic can be advantageous because:
- It provides flexibility.
- It gives options.
- It gives imagination.
- It is more forgiving.
- It allows for observation.

Fuzzy logic is being used extensively in consumer products where the input is provided by sensors rather than by people. It is believed that fuzzy logic will also become valuable in the next generation of computer systems. In a fuzzy logic system, the technology building blocks can be used in series or in parallel.

A fuzzy reasoning model has been constructed and applied to contract decision-making in Hong Kong. According to professional experience, six inputs are considered essential factors in determining type of contract (1) the scale of the project, (2) the type of work to be carried out, (3) the characteristics of the client, (4) the time constraints, (5) the source of materials for construction, and (6) the building design. These factors determine whether the contract should be in the form of a simple quotation, a lump sum contract with drawings and specifications, a schedule of rates, a management contract, a lump sum contract in standard form without quantities, or one with quantities. Many decision rules are then constructed based on expert opinions.

Automatic Programming

Automatic programming is a super-compiler. It is a program that can take in a very high-level description of what the program is to accomplish and produce the program in a specific programming language. One of the important contributions of research in automatic programming has been the notion of debugging as a problem-solving strategy. It is often much more efficient to produce an inexpensive solution to a programming or robot control problem that has errors, and then modify it than to insist on a defect-free first solution.

Natural Language Processing (NLP)

Natural language processing (NLP) allows computer users to communicate with the computer in their native language. This technology allows for a conversational interface, in contrast to using a programming language of computer jargon, syntax, and commands. There has been some limited success in natural language programming—current systems can only recognize and interpret written sentences. Although this ability can be used to great advantage with some applications, a general NLP system is not yet possible. There are two types of natural language processing:

1. *Natural language understanding* investigates ways to allow the computer to comprehend instructions given in ordinary English so that computers can understand people more easily.

2. *Natural language generation* allows computers to produce ordinary English language so that people can understand computers more easily.

Intelligent Agents

An intelligent agent is software that can perform intelligent functions. This is one of the fastest-growing areas of research and new application development on the Internet today. The key attributes of intelligent agents are:

- *Autonomy.* The intelligent agent must have the capability to take actions toward completing some tasks or objectives without impetus from the user—there must be an element of independence. Much like human agents, they take our direction, interests, wants, and desires as input and action to achieve the goal.

- *Communication ability.* Intelligent agents access information from the environment about its current "state" in the course of achieving their objectives. This requires that they be able to communicate with the repositories of this information.

- *Capability for cooperation.* An intelligent agent must have a collaborative "spirit," a willingness to work with other agents on complex problem domains that require multiagent efforts.

- *Capability for reasoning.* Intelligent agents should be able to reason based on a knowledge base that includes rule-based, case-based, and artificial evolution-based systems. Rule-based reasoning is using user preconditions to evaluate conditions in the external environment. Case-based reasoning draws suggestions or conclusions based on prior scenarios and the actions that resulted, from which it deduces future moves. Artificial evolution-based reasoning means that agents can acquire new knowledge each time problems are solved.

- *Adaptive behavior.* An intelligent agent must have a mechanism for accessing the current state of its external domain so that it can improve or readjust it for further actions.

INTELLIGENT COMPUTER-AIDED INSTRUCTION (ICAI)

Intelligent computer-aided instruction (ICAI) refers to machines that can tutor humans so that information can be passed on. Computer-assisted instruction (CAI), which has been in use for several decades, brought the power of the computer to bear on the business educational process. Now AI technologies are applied to CAI systems in an attempt to create computerized tutors that shape their teaching techniques to fit the learning patterns of the individual employee.

CHAPTER 3

Neural Networks

Expert systems typically require huge databases of information gathered from recognized experts in a given field. An ES will then ask the user questions and deduce an answer based on the responses given and the information in the database. These answers are not necessarily right but should be logical given the information provided.

Neural networks (NNs) are an evolving technology in which computers actually try to learn from the database and the operator determines what the *right* answer is. The system gets positive or negative responses from the operator and stores that data in order to make a better decision the next time. While still in its infancy, this technology shows promise especially for fraud detection, economic forecasting, and risk appraisals.

The idea behind NN software is to convert the order-taking computer into a thinking problem-solver. This would allow computers to take over some of the more mundane decision-making, like that of an accountant determining if a lease is an operating or capital one. NNs are thus software programs that simulate that aspect of human intelligence that learns from experience. For example, each time a NN makes the right decision (as predetermined by a human instructor) on recognizing a number or sequence-of-action pattern, the programmer reinforces the program with a confirmation message that is stored. If the decision is wrong, the message is negative. Thus, the NN gradually builds experimental knowledge in that subject.

Today most NNs take the form of mathematical simulations embedded in software that runs on ordinary microprocessors. In the future, network chips are likely to emerge that will dramatically increase both the speed of operations and their applications. These chips will be used to mimic decision operations and carry them out the way humans do.

BUSINESS NEURAL NETWORK APPLICATIONS

NN applications are numerous. Many potential users first learn about NNs by reading about how they are applied in predictions about financial markets. Several well-known investment groups claim that at least some of their technical analysis of financial markets and portfolio selection is performed using neural networks. Other successful NN applications are analysis of market research data and customer satisfaction, industrial process control, forecasting, and identifying credit card fraud.

When Mellon Bank, like a number of other banks, installed a NN credit card fraud detection system, the savings realized were expected to pay for the new system within six months. These systems can recognize fraudulent use based on past charge patterns far more accurately than other available methods. Currently, the neural network computer employed at the Mellon Bank Visa and MasterCard operation in Wilmington, Delaware, daily monitors 1.2 million accounts. One of its functions is to scan customer purchases for spending patterns that may indicate stolen credit cards. The NN compares purchases with customer behaviors. It also generates data without being told to do so, because the system has been programmed to take the initiative and think like a human.

One way NNs help accountants is with internal audits. Among the complex pattern recognition tasks NNs can perform are forecasting earnings and detecting fraud. The ability to forecast company earnings can be useful in planning an audit or helping management design an operating strategy.

While there are many types of NNs, those most useful to financial managers can be classified into four categories (1) prediction networks, (2) classification networks, (3) data filtering networks, and (4) optimization networks. NNs are much more tolerant of imperfections in input data than conventional computers, and more efficient in solving pattern recognition problems. NNs also overcome many ES limitations, including the necessity to extract knowledge from experts and the inability to learn. Ernst & Young in Dallas is working on an application that would allow financial managers to improve how they handle working capital.

Neural networks are particularly helpful in many business problems when information is not easy to quantify, as in portfolio management. A portfolio manager must continuously scan for nonperforming stocks, while on the other side stock analysts are looking for undervalued stocks. NNs are particularly good when deductive reasoning gives mixed results because they use inductive reasoning. There is a large store of historical information about good and bad investments that can be analyzed for relationships that may be quite subtle. Shearson-Lehman is using NNs to predict stock patterns.

INSOLVENCY PREDICTION

Neural networks can be useful in predicting bankruptcy, using tools already in place to improve prediction. For example, if the ratios chosen for the Altman's Z-score insolvency predition model are used but the NN is allowed to form its own conclusions, the predictive abilities of the Z-formula are greatly enhanced. This is of significant value to managers, creditors, and investors because misclassification, particularly of a firm that is going bankrupt, has huge monetary implications.

The *Neural Bankruptcy Prediction Program* developed in 1995 by Dorsey, Edmister, and Johnson shows how this works. This is DOS-based software that can be downloaded from the University of Missouri Business School's Internet Web page. The following gives some background and explains how this program can be applied.

Since Beaver (1966) first introduced the univariate statistical technique and showed how financial ratios could be used to predict the failure of a corporation years before it happened, many complex models have been formulated, among them Altman's multivariate analysis (1968), Edmister's multivariate analysis with trend factors (1972), and the recursive partitioning algorithm of Frydman, Altman, and Kao (1985).

Although the insolvency prediction models are highly complex and difficult to calculate, their use has become more practical now that we have advanced spreadsheet programs like Excel and Lotus, personal computers, and bankruptcy prediction software.

Insolvency prediction models and ratios are summarized in Table 3-1.

TABLE 3-1. Summary of Bankruptcy Prediction Models (Nonfinancial)

Year	1968	1966	1972	1985	1994
Analysis	Multivariate	Univariate	Multivariate with Trend Factors	Recursive Partitioning Algorithm	Artificial Neural System
Company size	Mid-Large	Large	Small	Large	Large
Accuracy in predicting insolvency	79%	87%	93%	96%	96%-100%
Ratios Used					
Cash flow/Total assets				√	√
Cash/Total sales				√	√
Cash flow/Total debt			√	√	√
Current assets/Current liabilities		√		√	√
Current assets/Total assets				√	√
Current assets/Total sales				√	√
Earning before tax and interest/Total assets	√			√	√
Retained earnings/Total assets	√			√	√
Net income/Total assets		√		√	√
Total debt/Total assets		√		√	√
Total sales/Total assets	√			√	√
Working capital/Sales				√	√
Working capital/Total assets	√	√		√	√
Quick assets/Total assets				√	√
Quick assets/Current liabilities			√	√	√
Quick assets/Total sales				√	√
Equity market value/Total capitalization				√	√
Income (total assets)				√	√
Income (interest+15)				√	√
Cash/Current liabilities			√		
Current liabilities/Equity			√		
Inventory/Sales			√		
Equity/Sales			√		
Equity market value/Total debt	√				
Income/Total capitalization					

With recent technological advances, the artificial neural network (ANN), a form of artificial intelligence, has emerged as a tool for predicting bankruptcy. As defined by Dorsey and his colleagues (1994), ANNs are "simplified models of the interconnections between cells of the brain." They can learn, generalize, and abstract like the human mind. Because of these unique abilities, the ANN is superior to previous models. It is:

- Free of distributional assumptions (because it learns the underlying functional relationship from the data itself).
- Free of collinearity.
- Able to approximate functions.
- Able to achieve greater accuracy in predictions.
- More robust.

Let's look at how the *Neural Network Bankruptcy Prediction Program* could have been used to predict 1995 bankruptcies from 1994 financial ratios. In the original study, Dorsey and his colleagues studied 1989-1991 bankruptcies.

Two sets of data were examined for this example:

1. Ten public companies each with assets of over $20 million that filed Chapter 11 bankruptcies in 1995 were chosen from the *Bankruptcy Data Source* published by New Generation Research.
2. Ten solvent companies each with assets of over $20 million were chosen on the basis of poor earnings and cash flows.

Financial data were obtained from America Online's "Financial Reports" database. Financial reports from 1994 (one year before the insolvency) were used. Then, 18 ratios required by the Neural Network Bankruptcy Prediction Program were calculated (see Table 3-2). The results are shown in Table 3-3.

Companies with values over 0.4 (the threshold value) are predicted as insolvent. Three of 10 solvent companies (as of early 1998) received values over 0.4, indicating a likelihood of bankruptcy. Although those companies did not go bankrupt in 1995, they can be considered high-risk companies. The analysis of the insolvent companies was impressive—all are recognized.

TABLE 3-2. Data Set Ratio Definitions

Ratio	Definition
CASH/TA	Cash/Total assets
CASH/TS	Cash/Net sales
CF/TD	Cash flow operations income/Total liabilities
CA/CL	Total current assets/Total current liabilities
CA/TA	Total current assets/Total assets
CA/TS	Total current assets/Net sales
EBIT/TA	(Interest expense + Income before tax)/Total assets
LOG(INT+15)	LOG(Interest expense + Income before tax)/Total assets + 15)
LOG(TA)	LOG(Total assets)
MVE/TK	Shareholders' equity/(Total assets - Total current liabilities)
NI/TA	Net income/Total assets
QA/CL	(Total current assets - Inventories)/Total current liabilities
QA/TA	(Total current assets - Inventories)/Total assets
QA/TS	(Total current assets - Inventories)/Net sales
RE/TA	Retained earnings/Total assets
TD/TA	Total liabilities/Total assets
WK/TA	(Total assets/Net sales)/(Working capital/Total assets)
WK/TS	1/(Net sales/Working capital)

TABLE 3-3. 1994 Data Analysis for Selected Firms

Year	Company	Bankrupt (1-yes/0-no)	Value	Insolvency Predicted (1-yes/0-no)	Error
1994	Apparel Ventures Inc.	0	0.38085095	0	
1994	Apple Computer Inc.	0	0.00267937	0	
1994	Biscayne Apparel Inc.	0	0.43153563	1	
1994	Epitope Inc.	0	0.18238553	0	
1994	Montgomery Ward Holding Co.	0	0.06268501	0	30%
1994	Schwerman Trucking Co.	0	0.57828138	1	
1994	Signal Apparel Co.	0	0.37081639	0	
1994	Southern Pacific Transportation	0	-0.3514775	0	
1994	Time Warner Inc.	0	0.57828138	1	
1994	Warner Insurance Services	0	-0.1020344	0	
1994	Baldwin Builders	1	0.57828138	1	
1994	Bradlees Inc.	1	1.00828347	1	
1994	Burlington Motor Holdings	1	1.00828347	1	
1994	Clothestime Inc.	1	0.43153563	1	
1994	Dow Corning	1	0.65517535	1	0%
1994	Edison Brothers Stores	1	0.61124180	1	
1994	Freymiller Trucking Co.	1	1.00828347	1	
1994	Lamonts Apparel Inc.	1	1.28955072	1	
1994	Plaid Clothing Group Inc.	1	1.28955072	1	
1994	Smith Corona Co.	1	0.43153563	1	

This analysis confirms that the Neural Network Bankruptcy Prediction Program is a reliable tool for screening financially distressed large companies. Despite the small sample size, 100 percent accuracy in predicting insolvency is remarkable. The program is a relatively simple and easy way to analyze data and interpret results. It holds promise even for those not proficient in mathematics.

Further, the program allows the user to import data from the Compact Disclosures™ CD, which contains 10K reports for all corporations and is available to financial analysts. The program calculates ratios after drawing data from the CD. This eliminates time and effort in collecting data and doing calculations. Its high accuracy rate may make the program useful as a decision-making tool for buying and selling bonds. It may also be valuable in screening large corporations or financial institutions when:

- Lending money.
- Contracting to supply or receive products and services.
- Leasing property.
- Buying stocks.

Because there are many financially distressed corporations and financial institutions in active operation, it is a good idea to analyze the likelihood of their going bankrupt before deciding to do business with them.

REFERENCES

Altman, E. I. "Financial Ratios, Discriminant Analysis and the Prediction of Corporate Bankruptcy," *Journal of Finance* (1968) 9: 589-609.

Beaver, W. H. "Financial Ratios as Predictors of Failure" *Empirical Research in Accounting: Selected Studies,* Chicago: University of Chicago Press, (1966): 71-111.

Dorsey, R.E., Edmister, R.O., and Johnson, J.D. *Bankruptcy Prediction Using Artificial Neural Systems,* Published on University of Missouri Business School's Web page, 1995.

Edmister, R. O. "An Empirical Test of Financial Ratio Analysis for Small Business Failure Prediction" *Journal of Financial and Quantitative Analysis,* (1972): 3: 1477-93.

Frydman, H., Altman, E.I., and Kao, D. "Introducing Recursive Partitioning for Financial Classification: The Case of Financial Distress," *Journal of Finance,* (1985): 40: 269-91.

Various issues, *PC AI Magazine,* Knowledge Technology, Inc. Available at: <*http://www.pcai.com/pcai/New_Home_Page/ai_info/neural_nets.html*>.

CHAPTER 4

Artificial Intelligence in Business Finance

The two main AI versions used in finance applications to date are expert systems and neural network technologies. Fuzzy logic, genetic algorithms, and chaotic models are still in the research phases for potential financial applications. However, ES and NN technologies respond well to the needs of those working in finance.

This chapter deals first with expert systems and second with neural network applications in finance.

ESs work well for finance applications because they are rule-based, and so are the granting of credit, insurance underwriting, insurance claims processing, establishing insurance reserves, selecting securities, consistent application of constraints in managing investment portfolios, hedge advising, real-time data feed, trading rules and generators, financial advice for multinational companies, 24-hour trading systems in international markets, and international arbitrage. They are also popular because most financial applications are already computerized. To the extent that computerized financial systems follow structured programming techniques, converting them to a knowledge-based system requires only an appropriate inference engine, a knowledge engineer, and an application expert.

BANKING: GRANTING AND MONITORING CREDIT

The ease of use, accuracy, and reliability of expert systems has improved the quality and increased the quantity of services banks can deliver to customers. The two main areas where banks use ESs are in making decisions about granting credit and in monitoring the performance of credit customers.

Credit Granting

The initial application for credit typically requires the potential customer to provide financial information, personal and employment information, credit references, and in some cases audited financial statements. The bank will also obtain information from a credit-reporting agency, which it will review and enter into a credit scoring system. Often the credit-scoring system was built using an ES, which can determine whether or not to grant credit to a potential customer and how much credit to offer initially (the credit limit). Rules for granting credit under different conditions can be established as part of the knowledge base of an ES, and the loan approval staff can invoke the inference engine as they process the application and review the information submitted. The inference engine can be designed to mirror the credit application form the branch manager uses to take an initial order from a potential customer.

Approving additional credit for an existing customer involves determining whether adding the new amount asked for to the customer's outstanding balance would exceed a reasonable credit limit for that customer. While the process appears simplistic, the decision process for rejecting the request can be very complex. Coding a series of decision rules into a knowledge base makes it possible to readily process many routine requests and situations where the credit limit can be extended (limit was established years ago and never revised, the account is current and the excess over the limit is within a reasonable range, additional compensating balances have been established since the last credit review, the customer has acquired new collateral assets, etc.). The rules can be applied relatively quickly with an ES. If the application for extension of credit is still unacceptable, it can be flagged for a loan officer's review before rejection.

Credit Monitoring

Banks and other financial institutions monitor credit particularly to identify nonperforming loans early and to spot fraudulent use of credit. Use of credit, data on payments, payment timing, and requests for new lines of credit involve many complex decisions and therefore many complex rules. By using an ES to capture in the knowledge base these rules of stress on credit performance and fraudulent use, the bank can automatically monitor all credit accounts easily.

Banks maintain other information besides credit information. They handle payroll accounts and general checking accounts. They can also use general and specific industry statistical databases to identify industry trends. For each credit customer, the ES can apply all this information to identify particular customers who are financially stressed and run the risk of potential default or credit failure. The expert system automatically performs either monthly, weekly, or in some cases daily statement runs. Accounts that appear to be threatened by financial stress are promptly flagged for a loan officer's review. This minimizes potential losses to the bank by alerting the loan officers to take action before the customer fails to pay.

Fraudulent use of credit often involves activities that the perpetrator believes will escape detection by bank personnel and systems. An ES can analyze complex relationships very quickly. Based on the type of credit, its intended use, and the customer's history, an ES can evaluate each transaction for fraud potential by searching the appropriate knowledge base. Transactions suspected of fraud can be flagged for immediate follow-up and the ES will make suggestions on what kind of follow-up activities the loan officer or security staff should do. That's how expert systems help to minimize bank losses on credit.

INSURANCE: UNDERWRITING, CLAIMS PROCESSING, AND RESERVING

The complex activities insurance companies engage in prove ideal for expert systems. There are plenty of databases for automatic retrieval of data and because insurance is a regulated industry, the rules for many of these activities are significant and complex. Expert systems are successfully used especially for insurance underwriting, claims processing, and reserving.

Underwriting

In insurance underwriting the company must decide whether a given risk is acceptable. Many insurance companies specialize in just one or a few types of insurance, such as life, casualty, health, or automobile insurance. These companies are always seeking new customers to increase the base over which to spread the risks, and for new customers, the companies process an application.

Insurance applications require the applicant to disclose considerable information, and the company will typically search other databases to confirm that information as well as supplement it. After verifying the information supplied, the insurance company assesses the insurability of the customer by assessing the risks associated with that particular coverage for that particular customer. The analysis is complex. That's why many insurance companies use ESs to help their underwriters make the call.

Knowledge engineers work with the underwriters to develop the decision rules that are entered into the ES knowledge base. When the inference engine is designed, the experts (the underwriters) and the engineers identify confidence factors for various situations based on the company's past experience with similar customers. After a customer submits data and it's verified, the ES reviews the application and makes recommendations to the underwriter about whether to accept or reject a particular customer and what riders should be included with a standard policy.

Expert systems have significantly enhanced underwriting by reducing the time needed to process an application for new insurance and by identifying potentially uninsurable risks early. When a problem risk is identified, the ES gives the insurance company time to get more information or to do more analyses for a given applicant.

Claims Processing

Claims range from the routine to the complex and questionable. Determining the amount payable under the insurance contract and whether or not the claim is fraudulent are two areas where expert systems are valuable to insurance companies.

Decision rules for determining the validity of an insurance claim can be identified and coded into the ES knowledge base by the claims processors and a knowledge engineer. When a claim is submitted, the ES inference engine queries the information from the claim, the claims adjuster, and different databases. The expert system then either makes a recommendation about paying the claim, requests additional information from the customer or others, or recommends whether or not to investigate. Recommendations on further investigation will outline where the ES takes issue with the data, so that investigators know where to start their investigation.

The benefits of using expert systems include rapid processing of routine claims, sophisticated assessment of complex claims, and an early warning system for possibly fraudulent claims. Insurance companies have also found that using expert systems improves the quality of claims processing.

Reserving

Insurance companies are regulated by the states. Each state has unique requirements for how much reserves the company must maintain and the rules are invariably complex. Reserves are the amounts companies must keep on hand to be sure that obligations to policyholders and third parties are covered. The more states a company operates in, the more complex the reserving rules it must honor. Past success of managing reserves for future claims against current payouts also affects reserve calculations. Expert systems are very effective at helping insurance companies manage their reserves.

The decision rules for maintaining reserves, including the requirements of each state, are established in the knowledge base of the expert system. Periodically (weekly or monthly), the ES will review new and canceled policies, claims, claims payments, market performance, and portfolio risk. The ES then makes recommendations to the reserve manager about what actions to take to maintain reserves within state regulations and flags any trends in payouts or special risks associated with new policies. The manager can then assess what decisions need to be made about the reserves.

PORTFOLIO MANAGEMENT

Much of portfolio management responds to complex decision rules. Portfolio managers trade securities, make recommendations about buying or selling securities, and provide hedging advice to the owners of the portfolios that they manage. Expert systems prove valuable in a host of portfolio management activities, among them selection of securities, consistent application of constraints, and providing advice.

Security Selection

Choosing securities for a portfolio requires a significant number of decisions relating to questions like the risk of the security, short and long-term growth considerations, current and expected performance in the marketplace, the goals and objectives of the portfolio, and owner or legal constraints, if any. Expert systems are very effective at assessing such decision factors quickly and making recommendation to help portfolio managers make prompt but well-founded decisions about which securities to include in the portfolio and which to get rid of.

Each of the decision rules associated with each factor is incorporated into the ES knowledge base. When a new security is being considered, the portfolio manager invokes the ES inference engine to query the knowledge base for each factor. After combining the results for each factor, the ES makes a recommendation about the security. It also expresses how confident the system is in the recommendation made and why that confidence level was selected. This gives the portfolio manager an exhaustive analysis of any security being considered.

Used in conjunction with a database of securities already in the portfolio, the expert system will analyze their characteristics to see if acquiring the new security will significantly alter the balance of the portfolio along each of the factors considered. Thus, a new security that is high risk may get a positive recommendation if the percentage of high-risk securities is currently low and a negative one if the percentage is currently high. This ability of expert systems to adapt to changing environments makes them ideal for helping managers to maintain the right balance of investment performance for a given portfolio.

Consistent Application of Constraints

Portfolio management decisions are often directed by the owner's preferences. This increases the complexity of the decisions to be made; if the manager handles more than one portfolio, he can fall into serious errors by overlooking a constraint or applying one that belongs to a different portfolio. Expert systems assure that the right owner's requirements are considered before the manager takes any action related to a security.

The owner's preferences and requirements are expressed as decision rules in the ES knowledge base. Each time an action needs to be taken, the portfolio manager can invoke the inference engine to determine whether the decision fits within the constraints set by the portfolio owner. Owner profiles can also be created for various securities that might potentially be included in the portfolio and incorporated into the ES for assessment of a given recommendation or action. The ES reviews a proposed action against the owner's profile and reports on how the owner is likely to react. The expert system may also express a confidence factor in association with any recommendation to guide the portfolio manager in making the ultimate decision about a security.

Expert systems enhance the quality of service managers provide to portfolio owners. The expert system remembers each owner's unique concerns and helps the portfolio manager avoid violating those constraints. In evaluating investment options, an ES quickly identifies problems, particularly with pooled funds. The ES gives the manager time to explore alternatives and improve the performance of each portfolio while honoring its unique requirements.

Hedge Advisor

Market hedges are important investments used to balance the performance of an entire portfolio or to manage risk. Decisions about whether to hedge with securities that are counter-cyclical to the rest of the portfolio or whether to use options are very complex. Managers use expert systems to monitor the performance of both the market and specific securities to see whether the hedge objective is being attained.

Expert systems track the performance of hedge securities relative to the entire market and identify securities that may be potential hedge instruments for a given portfolio. Decision rules about hedge securities are entered into the ES knowledge base. Daily or continuous data feed into an expert system allows it to query the knowledge base to make recommendations about hedge securities. Confidence factors enhance the recommendations the ES makes to the portfolio manager.

ADVICE ON TRADING

Trading advisors give customers valuable ideas about when to buy and when to sell specific securities. Given the sometimes wide-range of trading values throughout the day for certain securities and the need for rapid communication, these advisors find sophisticated computer processing of information crucial. Expert systems help them to assess real-time data on current market performance of securities, establish trading rules, generate new trading rules, and analyze past advice to customers.

Real-Time Data Feed

Securities quotes are offered instantly over the Internet and via direct feed continuously throughout the day. Because some prices swing rapidly, it's sometimes hard to find time to assess what's happening with those securities. The advisor must formulate a recommendation to buy, sell, or hold a given security very quickly. ES allows them to incorporate the data feed into a usable tool for providing fast and effective investment advice.

Decision rules about making recommendations to buy, hold, or sell a security are coded into the ES knowledge base. The system monitors the direct data feed from the market associated for a given security or group of securities; as prices change, the ES inference engine automatically queries the knowledge base to form both a recommendation about what to do with the security and a reasonable confidence factor in making that recommendation. If the advisor accepts the recommendation, the decision is automatically transmitted to the advisor's customers.

Expert systems empower trading advisors to provide up-to-the-minute advice to customers based on the latest direct market data feed. This provides real value in real time for investors.

Trading Rules and Rule Generators

Besides giving specific advice on what to do with specific securities, groups of securities, or classes of securities, trading advisors also give customers trading rules for accomplishing certain objectives. The rules are derived from trades that worked successfully to reach those objectives in the recent past. Since the market is constantly changing, the rules for trading to accomplish a particular objective also change constantly. ESs not only communicate the most recent rules but also what can be used to generate new trading rules.

Decisions about how to formulate a trading rule for a given objective are captured in the ES knowledge base. The rule-generating expert system monitors trading activity throughout the day in a given security, group, or class, making recommendations about new trading rules throughout the day. Those recommendations can be automatically incorporated into the trading software and communicated to the customer as soon as the recommendation is formulated, if the advisor so decides after reviewing the rule recommendation and the confidence factor associated with it.

ES rule generators empower trading advisors to formulate and transmit new trading rules to customers in real time as the market is adjusting to changing circumstances.

Critics

The true test of any investment advice is to subject it to critical review and analysis. The number of factors that affect securities trading are extremely complex and difficult to apply across a broad range of securities. Advice about one security is relatively easy to review critically. ESs make it possible to quickly, accurately, and consistently assess the results of advice offered to customers and identify solid decision criteria and any questionable hypotheses used by the advisor.

The advisor sets the criteria for determining when an outcome of applying trading advice and trading rules is successful. Decision rules that incorporate success criteria are in turn incorporated into the ES knowledge base. A critic expert system then monitors the performance of securities and makes an observation about which trading advice or rules succeeded and which did not. The confidence factors associated with the recommendations guide the trading advisor in adjusting the methods and hypotheses used to formulate trading advice.

GLOBAL FINANCIAL MARKETS

The fact that financial markets today operate globally presents unique challenges to those in the industry, among them the uniqueness of individual company legal and regulatory reporting, the interdependence of economies, round-the-clock operation of trading markets, the need for instruments to hedge the risks of a variety of local markets, currency differentials, and opportunities for arbitrage. Expert systems can support investors and their advisors in all these areas. They can advise multinational corporations on their financial statements, manage 24-hour trading programs, make recommendations on hedge transactions, and identify arbitrage opportunities.

Financial Statement Advice

Multinational firms must conform to a vast number of financial disclosures and laws that differ significantly from those in their home countries. The many inconsistencies between national and local laws can be overwhelming. ESs help multinational financial professionals to identify differences and suggest what the firm must do to comply with the laws of countries in which the firm operates so as to minimize tax payments and avoid fines and penalties.

Details of the laws of the countries in which the firm does business are coded into the ES knowledge base. Each time a transaction is executed in a given country, the inference engine checks to determine the laws there. The expert system also checks decision criteria for the home country. It then recommends a specific treatment for a given transaction to optimize the results to the firm within the laws of both the home country and the country where the transaction takes place. This saves multinationals significant time in preparing financial disclosures and by reducing costs for taxes and penalties, brings more money to the bottom line.

Twenty-Four-Hour Trading Programs

The many investment opportunities in financial markets around the world may not be available on an investor's home market; a significant number of portfolios contain securities that are traded only in single markets that may be in Asia, Europe, Africa, as well as the U.S. To effectively manage portfolios, trading firms would have to staff their operations 24 hours per day if it weren't for expert systems. ESs can monitor market activity and issue buy and sell orders independent of a human being. These traderless ESs allow smaller firms to participate effectively in international markets without the need for 24-hour staffing.

After an expert system is tested at the start of each day, the firm can check trades the system made during the previous night. At the end of their workday, staff can adjust the expert system to take into account what happened that day in their home country. As soon as a foreign market opens, the ES is monitoring it to determine which securities to buy and which to sell. Controls can be entered so that the computer calls a trader who is on stand-by if any emergency arises in the foreign market, such as a run on a security, a market crash, or an unexpected acceleration of price either in a monitored security or the market as a whole.

Hedges

The uncertainties and risks associated with international financial markets means that investors both need and have significant opportunities for taking hedge positions. Hedges may include interest rate swaps, currency swaps, options, and futures contracts. Tracking trends and short-term movements of the markets in hedges is an important aspect of financial management; expert systems provide real, immediate help in managing and monitoring them.

To do this, the financial manager and a knowledge engineer once again establish decision criteria for the ES knowledge base. The ES inference engine continuously monitors data about hedges. When conditions in the international market change or when action is needed to protect the hedge, the ES issues recommendations. These are important tools for the international financial manager concerned to properly manage hedge positions.

Arbitrage

International financial activities, instruments, and transactions offer a significant number of opportunities for profiting from arbitrage. In a typical arbitrage arrangement, a packaged contract is executed at an effective rate of, say 10 percent, based on the blended risk associated with the package. The buyer then breaks up the package, selling 90 percent of it at an effective rate of 8 percent. The buyer continues to collect 10 percent on the entire package and pays the buyer of the 90 percent portion only 8 percent. The 2 percent difference on the sold portion constitutes the arbitrage fee the initial buyer enjoys. Calculated as a return, the arbitrageur is earning an effective rate of 28 percent. In international trade, bundling and unbundling of goods, services, financial instruments, and other contracts is fairly common. Not all unbundling activities offer arbitrage but many do. This provides important opportunities to enhance profitability in international markets.

The potential conditions for arbitrage arrangements can be determined and entered into an ES knowledge base. The expert system can then either be run regularly or invoked when certain conditions arise; once operating it determines the potential for realizing arbitrage profits in an international transaction or market place.

NEURAL NETWORKS IN BUSINESS FINANCE

Neural networks are applied in banking, insurance, portfolio management, real estate brokerage, investment banking, and retail brokerage. For each application we will demonstrate how to plan the NN, identify and select data, train the NN, use it, retrain it, follow-up procedures, and any available commercial software packages and services.

The superior pattern recognition power of neural networks and the speed at which they operate offer significant opportunities for their use in today's information technology (IT) working environments. NNs are used in finance to assist in decision-making, provide early warning signals, and confirm assessments.

Some neural networks have been effectively integrated into other software applications to enhance data, speed up its processing, or make decision recommendations. In these applications, the NN may be a front-end, middle, or back-end process. Some institutions have found that NNs can rapidly isolate conditions in original data (front-end processing) and signal the ES knowledge base (or other decision support submodule) appropriate for a given circumstance, reducing the time of access of the search engine; this is known as pattern recognition. In middle processing, the NN may take the results of a front-end process and signal the appropriate subsequent process; this is called filtering. In back-end processing, the NN uses input from other systems to make a final decision, the prediction. Finance activities regularly use all three functions to enhance their results. Neural networks may even be used in analysis of mergers and acquisitions.

Banking

The general activities of banks, savings and loans, and credit unions include among others demand deposit services, savings services, lending, and trusteeship. Most of these services require a large number of computer transactions, thus offering significant opportunities for data analysis by neural networks.

Bankers must often make decisions that have an impact on the success of the bank. These decisions include credit-granting decisions, detection of fraud, and fiduciary responsibilities under law or by trust agreement. This section explores how NNs can help bankers perform all these activities. Each section explores the steps associated with applying NNs in a given banking activity.

Granting Credit

Banks classify loans or credit arrangements into categories like credit cards, personal loans, commercial loans, and mortgages. When bank customers apply for credit, the bank assesses their creditworthiness and makes a decision about whether to extend the credit or deny the application. The funds are transferred. The loan is serviced. The loan is repaid.

Opportunities for using NNs in deciding whether to grant credit and how much include prescreening loan applications to eliminate obviously unworthy applicants, setting particular loan conditions, making credit recommendations, and monitoring debt service. Through prescreening and early rejection of bad loan candidates, the bank reduces processing costs and improves customer service by giving them immediate feedback.

Using NNs reduces the time required to perform analyses or provide reassurance about manual determinations of creditworthiness. In monitoring debt service, NNs can give early warning of troubled debt customers, opening up more opportunities for the bank to minimize losses.

Many banks have used NNs successfully to pre-screen loan applications on-line. Prescreening is conducted at branches by providing a trained NN part of the software available to branch loan officers.

When banks design NNs, they give them enough data to recognize patterns. Once the NNs are trained, they become part of other software that provides the user interfaces. The banks regularly retrain networks and update the software, then follow-up by monitoring how the networks perform.

Planning the NN for granting credit. An effective NN for granting credit needs a clear, precise definition of what the network should accomplish, when it should perform the task, and how the results should be communicated.

If the goal is to prescreen credit applications, decisions about data, NN type, and the outcome can be stated succinctly. For example, in granting a credit card to an applicant, the bank uses an application form that the applicant completes and returns to the bank, where the information on the application is entered into a computer. The computer then accesses credit bureau reports and other data within the bank and makes a determination about the creditworthiness of the applicant. The bank pays the cost of processing the application, including a fee to the credit bureau. If there is enough data on the application form to make an initial determination about creditworthiness, the bank can avoid such processing costs as credit bureau fees by using a neural network.

If the goal is to use the NN to help create the creditworthiness profile, the bank incorporates into the NN procedures for using the application and other data, including credit bureau reports and other bank records on the history of the customer (debt service records from personal loans or mortgage loans). When there are a large number of data sources, there are many relationships among the data and a neural network can accurately identify many of these relationships.

Identifying and selecting data. Applications for credit vary significantly—they can be the stub of a postcard or they can run into several pages of data with supplementary financial statements attached. In general, the more data available, the easier the job of the neural network. However, a more limited data set can sometimes yield good results. Generally speaking, the bank requires data from both successful credit experiences and failed credit experiences to adequately train a neural network.

The NN learns by identifying differences in patterns between successful loan applications and those applications that were approved but the loan later went bad. If the bank has no data on unsuccessful loans, it will be impossible to train a neural network. Some banks run several trials or pilot projects using a small sample (100 to 150) of customer accounts to identify the best combination of data. The pilot projects vary the data and number of inputs to find the best combination. In general, a pilot should include as much data as is available over the most recent history from the last two quarters to the last five years. The date of application may be important to the neural network because of the economic conditions at the time of the application, which might affect the applicant's creditworthiness.

Preparing the data may take a significant effort, because NNs require numerical data representation. Hence, codes that use alphabetical characters must be converted to an ordinal scale. The scale should have wide enough gaps to allow discrimination between observations (for instance, if there are four geographical regions, instead of numbering them one, two, three, and four, it would be better to use two, four, six, and eight).

Training the NN application. There are two possible approaches to training the neural network (1) a massive set (all loan accounts and all credit experience) or (2) a simplified set (a representative sample of loan accounts). With a massive set, the NN will need significantly more time for training, because training means reexamining the data hundreds of thousands of times. If the number of observations is large (tens of thousands or millions) the training iterations become millions or billions.

The risk of using smaller training sets, however, is that the NN will memorize the actual training data and then be unable to make output decisions that generalize when a broader set of input observations are presented to it. In any case, part of the training of a neural network is to hold out a sample of both creditworthy and uncreditworthy applications. This holdout sample is used to test the trained network with data it has not yet seen to see if it makes accurate decisions. This test assures that the NN can accurately distinguish between applications that are creditworthy and those that are not.

Using the NN. Once the NN is trained, it can be installed as a subroutine within any of the operational software programs the bank uses to process credit applications. If the goal is to prescreen applications, the NN can be executed at the end of the initial data entry program or at the end of input just after all the validity check routines have been executed. If this is done, the customer can be notified immediately if applying for credit at a branch or on-line. For applications that are processed in bulk through mail-in applications, rejected applications can be returned immediately.

If the goal is to help assess the general creditworthiness of a customer or mortgage collateral, the trained NN is embedded in the software near the end of application processing, after all inputs have been obtained. In most situations, the NN makes a recommendation to the loan officer or underwriter, who then makes the final determination based on professional judgment. In some cases, however, the NN makes the final decision about granting credit.

Retraining the NN. Over time, circumstances change and so do creditworthiness profiles. As the bank gains more experience with loans and losses, adding new loan experiences to the training set can retrain the neural network. Most banks that have a neural network retrain it every month or every quarter to reflect the most recent experience. Other banks delete the oldest data and replace it with the newest. The disadvantage of this approach is that the bank loses experience that it gained before using the neural network.

Follow-up and monitoring. Banks using neural networks should try to determine why loans fail and whether the network is failing to detect non-creditworthy applicants. If a network is only retrained once a year and if the rate of non-performing loans increases significantly during a particular month or quarter, the bank monitoring should indicate that retraining should be done more often.

Commercial software packages and services. There are any number of NN software programs that can be installed in a credit-granting institution. HNC Software Inc. offers *Capstone™ Decision Manager,* which can easily be integrated into an existing lending decision environment in the loan origination or application processing system. HNC also offers *Capstone Online,* an HNC-hosted ASP solution that offers automatic credit-decision-making. The software connects with point-of-sale machines as well as major consumer credit and business reporting agencies.

Early in 1999 HNC Financial Solutions introduced a product, *Gemini Verify Score,* a neural network-based package currently used by Equifax. This software is a neural net developed using the credit history database of the entire United States. It was pilot-tested at Branch Banking and Trust Company of Winston-Salem, North Carolina. The bank claimed the tool allowed the bank to enter into new markets without a major expansion of resources. Among the advantages cited were integration with other Equifax products and databases, faster review and approval time, approval of valid accounts without costly application verifications or investigations, reduction in the number of applicants who drop out of the verification process, and reduced opportunity costs for acquiring new accounts.

Neuristics, LLC offers *The Edge™,* a system that promises to identify good credit risks that have been traditionally misclassified by standard credit scoring tools. The software is primarily designed for identifying prospects in the subprime market; it can be integrated into existing bank credit software. The company also offers *Nexcore™* as the next generation of NN-based credit risk scoring. It is designed for all types of credit scoring, not just subprime.

Credit and Debit Card Fraud

Individuals working to steal money often try to defraud banks. Neural networks can help detect fraud in three areas (1) fraudulent financial disclosures as part of loan applications, (2) creditor debit card fraud, and (3) check fraud. They help banks detect fraud because they can recognize patterns in complex data that traditional models do not recognize.

One of the most common forms of bank fraud is theft of and subsequent unauthorized use of credit or debit cards. As information technology becomes more prevalent, more and more merchants throughout the world are using online authorization of credit and debit card purchases. Banks that have issued the cards receive notice of card transactions immediately. Using a neural network to monitor card transactions helps to immediately identify cards that may have been stolen and give banks an opportunity to refuse authorization of purchases and perhaps even recover lost or stolen cards.

Planning the NN. For fraud applications, the neural net should be designed to recognize patterns of card use that suggest that a card may have been stolen. A thief is generally aware that at some point the card owner will discover that the card is missing and report the loss to the bank. Therefore, a thief will attempt to charge up the entire credit limit of a stolen card in a short period of time. Usually, the thief will obtain goods that can be readily resold. This results in subtle changes in patterns of purchases (timing and types of goods purchased) that NNs have been successful at detecting. As card authorizations are called in or received online, the bank's IT software passes the account information through a neural network to flag patterns that indicate a thief is using a stolen card.

Once the bank's computer alerts a telephone operator who calls the merchant, the operator can question the person presenting the card and if necessary ask the merchant to seize the card and return it to the bank. Many cards have been recovered thanks to a neural network.

Identifying and selecting data. The data used by banks to build these types of neural networks include a number of observations such as past purchasing trends of the account, frequency of transactions, size of each transaction, characteristics of the customer (from the original application), and other account data. The bank should keep data from accounts subjected to fraud with similar non-fraud data. Banks need as much data as possible for this type of neural network, usually two to five years of history.

Training the neural network. Training a neural network to detect fraud is just like training it to grant credit (see pp. 63). Using a massive training set of all credit and debit card transactions takes significantly more time, but using a simplified set runs the risk of the NN not being able to generalize on broader sets of data. Again, it's important to use a holdout sample to test the training results on. This test provides assurance about the NN's ability to accurately distinguish between requests for authorization of legitimate purchases and requests that may result from fraudulent use of a credit or debit card.

Using the NN for fraud detection. Once the neural network is trained, the trained network can be installed as a subroutine within any of the bank's operational software. If the goal is to flag potential fraudulent card use, the NN is embedded in the software near the end of authorization processing after all the inputs have been obtained. Usually, the NN generates a message that is sent to a customer service representative for follow-up with the customer by phone as soon as possible.

In some cases, the NN is not only used to help prevent fraud but also to recover a stolen card. In these situations, the NN is embedded within the software that is used to authorize the credit card purchase. When a pattern emerges to indicate a potential fraudulent purchase, the software notifies the merchant seeking authorization to confiscate the card and contact the bank. Meanwhile the bank attempts to contact the legitimate owner of the card.

Retraining the NN. As circumstances change over time, so do patterns of fraudulent credit card use. As the bank gains more experience with credit and debit card fraud losses, adding new experiences to the training set can retrain the neural network. Most banks using NNs for this purpose retrain the network every month or every quarter to reflect the most recent changes in experience. Deleting the oldest set of data and replacing it with the newest, as some banks do, means that the bank loses all the experience gained before using the neural network.

Follow-up and monitoring. Banks using neural networks should be constantly monitoring them to see how well they are detecting fraudulent use of credit and debit cards. If the rate of undetected fraudulent use increases significantly during any month or quarter, the bank's follow-up procedures should mandate immediate retraining.

Commercial software packages and services. CyberSource Corporation and Visa U.S.A. collaborated on a neural network using member-reported data to detect credit card fraud. Their Cyber Source Internet Fraud Screen is used to detect e-commerce credit card fraud.

InterCept Group, Inc. offers community financial institutions The Bankers Bank Secure Debit Card Program, which uses a neural network to protect particularly against debit card losses.

ACI Worldwide offers several products marketed under the brands *OCM24®, TRANS24®,* and *i24™*. These products, which facilitate e-payments, incorporate a neural network-based e-fraud detection and management service for both card issuers and merchants.

Nester, Inc.'s PRISM CardAlert is a neural network fraud risk management system that banks have been using since 1992 to detect fraudulent card transactions. The software has been regularly praised for reducing fraud losses.

HNC Software Inc. offers the *Falcon™* and *Falcon™ Model 2000* fraud detection packages; they are used by a number of financial institutions to detect fraudulent credit card transactions. The company's *Falcon™ Debit* is a neural network-based fraud detection software package particularly designed for use with debit cards. Visa U.S.A uses it to monitor member debit card transactions. The HNC *Credit Intelligence Solution* is a comprehensive NN-based system to minimize credit risk, reduce fraud, and improve customer service.

HNC *Financial Solutions* offers Canadian banks a product similar to its Gemini Verify Score called the *Gemini Application Fraud Detector.* A number of institutions are currently using or testing its neural networks. Advantages cited include an analytical approach to fraud control and better identification of fraud losses, which are sometimes classified and written off as credit default.

Mellon Bank's Visa and Master Card operations daily monitor their accounts by scanning customer purchases and examining spending patterns that may indicate stolen credit cards.

TransAlliance LLP offers *Cardholder Risk Identification Service,* a neural network used by many credit unions to detect fraudulent use of debit cards. In addition to early detection and prevention of fraud losses, users claim reduced insurance premiums.

TRW's Discovery screens for possible consumer credit card fraud by looking for recurring patterns in user inquiry data.

Security Pacific's Debit Card Fraud Project detects and safeguards against fraud in the use of automatic teller machines.

Bank Check Fraud

A common form of bank fraud is the passing of bad checks. It's estimated that check fraud costs banks over $1 billion annually. As it becomes more prevalent, more and more merchants worldwide are using online authorization of check transactions. Using a neural network to monitor check transactions helps to identify potentially bad checks immediately and gives banks an opportunity to refuse authorization of the check transaction.

Planning the NN. For fraud applications, the NN should be designed to recognize patterns of use of bad checks. A thief, who is usually aware that at some point the merchant or bank will discover that the check is bad, will attempt to write checks for as many purchases as possible as quickly as possible. Usually, the thief will obtain goods that can be readily resold. These actions result in subtle changes in patterns of purchases (timing and types) that NNs have been successful at detecting. As check authorizations are called in or received online, the bank's software passes the account information through a neural network to flag patterns that suggest fraudulent checks. Banks then alert a telephone operator, who notifies the merchant immediately that the check authorization has been declined.

Identifying and selecting data. Banks use a variety of data in these types of neural networks—observations about past purchasing trends of the account, frequency of transactions, size of each transaction, characteristics of the customer (from the original application), and other account data. The bank should keep data from accounts that have been subjected to check fraud with good-check data, and incorporate two to five years of history.

Training the NN. Training a neural network to detect bad checks is just like training it to grant credit (see pp. 63). Using a massive training set of all credit and debit card transactions takes significantly more time, but using a simplified set runs the risk of the NN not being able to generalize on broader sets of data. Again, it's important to use a holdout sample to test the training results on. This test provides assurance about the NN's ability to accurately distinguish between requests for authorization of legitimate purchases and requests that may result from fraudulent use of a credit or debit card.

Using the NN to detect check fraud. The trained NN can be installed as a subroutine within any of the software the bank uses to process check authorizations. In most situations, the neural network sends a message to a customer service representative, who follows up by contacting the customer by phone as soon as possible.

The NN can sometimes be used to actually recover stolen checks. When the NN embedded within the software used to authorize the check purchase recognizes a pattern suggesting a possible fraudulent purchase, the software notifies the merchant seeking authorization to notify the authorities to arrest the perpetrator.

Retraining the NN. As the bank gains more experience with check fraud losses, adding new experiences to the training set can retrain the neural network. It's best to do this monthly or quarterly to reflect the most recent changes in experience.

Follow-up and monitoring. Banks using neural networks should constantly check to see how well they are detecting check fraud. Whenever the rate of undetected fraudulent check use is seen to increase significantly, the bank should follow up immediately with NN retraining.

Commercial software packages and services. HNC Software, Inc. offers *Falcon Cheque™*, a check fraud detection package that easily integrates with in-stream, off-line batch and real-time processing paths of existing check authorization software.

TeleCheck® Services, Inc. partnered with Deep View Systems to implement a neural network-based system to approve all check transactions. The network is built into the software TeleCheck uses in its check approval services.

Fiduciary Responsibilities

Banks execute their fiduciary responsibilities managing employing assets according to trust agreements. In these situations, the bank often functions as an investor for the trust funds. The bank must monitor the securities markets to attempt to optimize the returns on assets entrusted to them under the terms of a large variety of trust agreements. Neural networks can be used to identify specific securities from among those that are likely to yield the highest return. The neural network selects those securities that might help the trustee accomplish the goals of the trust agreement. It can assess the potential of a given security to yield targeted returns or meet other investment objectives. The need to assess both individual securities and what is happening in the various securities' markets is the main reason banks use neural networks to successfully execute the trustee function.

Planning the NN application. The design of the neural network varies slightly depending on whether it is analyzing individual securities or an entire market segment, though the procedures for executing the network are similar. To assess a single investment opportunity, the NN has to be able to identify the attributes of interest to a given trust portfolio. The trustee may infer the objectives or use indirect measures to achieve the investment objectives, but must design the NN to meet a defined objective like maximizing dividend income while preserving growth of the trust corpus. Definition is crucial to the success of a neural network in support of the trustee function. The plan should identify the required NN outcome and identify general data requirements.

Identifying and selecting data. The trustee locates sources for the data needed to meet the trust goals. These sources vary depending on the bank, but generally involve a combination of internal and subscription databases. Data include securities information, market segment or market data from past experience at the bank and beyond, outcomes related to accomplishing objectives (i.e., higher than average returns for a group of utilities securities). The bank should review additional databases from subscription services for potential inclusion in the input data set.

Training the NN to meet fiduciary responsibilities. If the bank assesses individual securities, results of past assessments are the best way to train a neural network to assess individual securities in the future. Otherwise, a fiduciary support NN is trained the same way other networks are trained (see pp. 63). The massive set in this case would be all trust funds investment experience and the simplified set would be the investment experience of a representative sample of trust funds; as always, a holdout sample is crucial.

Using the NN in fiduciary functions. As with other NNs, the trained network can be integrated with other decision support systems to monitor movements in capital markets. The resulting decision support system initiates buy, sell, or hold decisions with respect to the securities the bank has in trust. By monitoring changes in the market place, the NN can predict upswings and downswings in the market, signaling buy and sell decisions instantaneously.

Retraining the NN. As the bank gains more experience with the securities markets, adding new experiences to the training set can retrain the network. There is usually enough movement in the markets and the economy to justify at least quarterly if not monthly network retraining.

Follow-up and monitoring. Banks using NNs in making decisions about buying and selling securities have plenty of prompt feedback about the validity of those decisions in every morning paper. Follow-up procedures to retrain the network should be initiated immediately if the performance of securities is fluctuating noticeably from what the NN predicted.

Commercial software packages. Stock 100, Inc., a financial engineering consulting firm, uses NN-based proprietary software to advise financial institutions. Stock 100, Inc. offers on-line monitoring and trading in conjunction with CyBerCorp, Inc. to provide continuous real time services.

MathWorks, Inc. offers *Datafeed Toolbox,* which connects with Bloomberg L.P. financial information services to download data to *MATLAB®,* which incorporates a neural network, time series, and statistical functions into its securities analysis.

State Street Bank has filed a patent on a portfolio construction system developed around a neural network by State Street Global Advisors.

Insurance Companies

Insurance company financial activities where neural networks can provide particular value include underwriting, detecting fraudulent claims, and establishing and maintaining reserves. Because underwriting and claims processing involve a large number of computer transactions, they offer significant opportunities for NNs to handle data analysis. Insurance companies are constantly making decisions that affect their financial success as they decide whether to accept applications for insurance, detect fraudulent claims, and keep their reserves up to legal requirements.

Underwriting

Insurance companies classify applications into a variety of insurance plans based on the type of insurance requested. A neural network can prescreen applications to identify applicants who are poor risks at the very beginning of the process and again near the end, using actuarial tables, industry information, and personal and locality data to supplement the information available on the original application. Underwriters have to analyze the data accumulated and decide whether or not to offer insurance coverage; NNs can do most of the hard work.

By prescreening and rejecting unworthy applicants early, the insurance company reduces processing costs and improves customer service by giving customers immediate feedback. Using NNs in underwriting vastly reduces the time required to perform an analysis and can also validate a previous manual determination of insurability.

Prescreening Applications

The goal of using NNs in processing policy applications is both to reduce costs and staff time and to improve the quality of the underwriting so as to minimize losses. Many insurance companies use NNs successfully for on-line pre-screening of policy applications at the offices of their agents by providing a trained NN as part of the software made available to agents.

The carriers generally give their NNs enough data to recognize patterns of insurability or noninsurability, and then implement them as part of other software with user-friendly interfaces.

Planning the prescreening NN. Good NN design relies on a clear definition of what the network should accomplish, when, and how the results should be communicated.

For prescreening policy applications, the necessary decisions can be stated succinctly. If there is enough data in the application to make an initial determination of insurability, the insurance company can avoid additional subscription service bureau processing costs by using a neural network.

If the goal is to use the NN to help create an insurability profile, the carrier uses applications, its own data, and other data including service bureau reports and the records of other insurance companies on the history of particular customers (types of coverage, claims, and premium payment history). With such a large variety of data sources, NNs can work fast to accurately identify crucial relationships in the sea of information.

Identifying and selecting data. Generally, the insurance company needs to inform the NN using data from both successful and failed underwriting experiences. The network learns by being shown the applications for insurance that were successful and applications that were erroneously approved. If the insurance company has no data on failed policies, it will be impossible to train a neural network.

It's best to run several trials or pilot projects using a small sample of policies (100 to 150) and varying the number of inputs to the network so the insurance company can identify the best combinations of inputs for the NN. The ideal is to include as much data available. Even the date on the application may be important in helping the NN recognize certain general economic conditions that might affect the wisdom of a decision to insure an applicant.

Preparing the data may take a significant effort because NNs require numerical data representation, so coded alphabetical characters must be converted to either a cardinal or ordinal scale. In applying a cardinal scale, leave enough gaps to allow discrimination between observations (i.e., if there are mortality rates for four different geographical regions, number them, say, two, four, six, and eight rather than one, two, three, and four).

Training the NN policy-screening application. A neural network used for underwriting is trained just as any other NN is (see pp. 63).

Using the NN for prescreening applications. The trained NN can be installed as a subroutine within any of the operational software used to process policy applications. If the NN can be executed at the end of the data entry program or at the end of the input just after all the validity check routines have been executed, the customer can be notified immediately if the policy is denied.

If the idea is to assess the insurability of a customer or the underwriting process more broadly, the NN is embedded in the software near the end of processing the application, after all inputs have been obtained. Usually, the NN makes a recommendation to the underwriter about the application, who uses professional judgment to make the final determination. However, sometimes, the NN report is the only consideration in determining whether to issue a policy, especially when the coverage amount is low.

Retraining the NN. As the insurance company gains more experience with policy losses, adding new experience to the training set can retrain the neural network. Most insurance companies using NN retrain them every month or at least every quarter to reflect the most recent changes in payout experiences.

Follow-up and monitoring. Insurance companies using NNs for underwriting should analyze why policies incur losses and determine whether or not the NN is failing to detect uninsurable applicants (there may be independent external influences on payout patterns). As soon as a detection problem is identified, the network should be retrained.

Commercial software and services. HNC Software, Inc. sells *Capstone Decision Manager* for Insurance Underwriting. The software was designed primarily for property and casualty rather than life and health insurers. The software easily integrates with existing policy management systems and distribution channels.

Insurance Claim Fraud

All too often, individuals attempt to obtain money from insurers fraudulently by submitting fictitious claims, exaggerated claims, or claims for losses not incurred by the insured. False claims are the most common form of fraud on insurance companies. A neural network can monitor insurance claims to identify potential fraudulent losses and give the insurance company an opportunity to refuse authorization of payments or flag claims for further investigation.

Planning the NN application. For detecting fraud, the neural network should be able to recognize patterns of data asserted in claims that indicate the possibility that they are fraudulent. Because they know the insurance company is likely at some point to challenge the claim, a thief will try to make the data sound convincing, but data from false claims often has a pattern that, though subtle, can be recognized by a neural network. This type of NN can be embedded in the carrier's initial claims processing software. Some companies use NNs to alert a claims adjuster to the potential of fraud to flag the claim for further investigation.

Identifying and selecting data. For this application, the NN must be programmed with such observations as past claims, frequency of claims, size of each claim, characteristics of customers (from the original application), and other account data. The insurance company should keep data from all accounts that have been subjected to fraud.

Training the NN. A neural network used for fraud detection is trained just like any other NN (see pp. 63).

Using the NN for fraud detection. To flag potentially fraudulent policy claims, the NN is embedded in the software near the end of the authorization process, after all inputs have been obtained. Usually, the NN simply generates a message that is sent to a claims adjuster for additional investigation.

Retraining the NN. As the company gains more experience with fraudulent policy claims, adding new experience to the training set retrains the neural network.

Follow-up and monitoring. New experience should not just be added to the NN in retraining, it should be analyzed to make sure the NN is doing a good job in detecting fraudulent policy claims. If not, the network may need to be modified as well as retrained.

Commercial software packages and services. HNC Software's *AUTOADVISOR™* is a neural network-based software package that reviews auto personal injury policy claims for fraud. It looks at bundling and unbundling, duplicate charges, non-accident-related prescriptions, and policy overutilization.

The same company's *COMPADVISOR™* performs similar services in detecting fraud and abuse of workers compensation claims but can also be used to analyze auto personal injury claims.

HNC's *Complex Bill Service and Mail-in Bill Review Service* allows its experts to process claims for insurance companies using its own NN-based software.

Establishing Reserves

Regulators require insurance companies to keep on hand a percentage of their income as a reserve against a run of claims. Companies execute their reserve maintenance responsibilities by estimating the claims payouts expected over a given period, setting aside assets in reserve funds, and managing those funds as required by the law of each state where the company is incorporated or doing business. Because the insurance company functions as an investor of the assets that compose the reserve funds (as is a bank with trust funds), the company must monitor the securities' markets to optimize its returns.

The insurance company needs to accurately predict how much it should reserve to pay claims, taking into account the possible return it may be able to get on the invested funds in the market. NNs can be used to accurately predict claim payouts and to identify securities that will yield the necessary returns within the legal rules the company must be responsive to.

Planning the NN application. Predicting claims payouts and selecting the right investments require slightly different approaches to the design of the NN but the procedures for executing it are similar. For predicting claims payouts, the NN must know both past experience and any expected changes in the number of policies, coverages, and policyholder demographics. For assessing investment opportunities, the NN must be designed to identify those attributes of securities that relate directly to the objectives of the company's investment plan.

Identifying and selecting data. Data sources for a neural network reserving application will vary depending on the insurance company but will generally be a combination of company internal databases and databases from subscription services. Data selected for predicting claims payouts includes past experience with policyholders, actuarial tables, and profiles of past and current policyholders among other relevant data. Data selected for investment analysis includes type of security, market segment, or market data from past company experience and outcomes required relative to accomplish reserving objectives

Training the reserving NN. For the securities analysis aspect of a neural network, results of past assessments of securities are the best way to train it to predict what individual securities will earn in the future; this may require use of databases provided by subscription services as well as internal sources. Otherwise, a neural network used for reserving functions is trained just like any other NN (see pp. 63).

Using the NN for the reserving function. The trained NN can be integrated with other decision support systems to monitor both claims payouts and movements in capital markets. It can be asked to make initial buy, sell, or hold decisions on securities held as reserve investments. By monitoring changes in the market place, the NN can instantaneously predict the upswings and downswings in the market that are known to signal buy and sell decisions and revise estimates of how much to reserve for claims payouts.

Retraining the NN. As the insurance company gains more experience with claims payout and securities' markets, adding new experiences to the training set can retrain the neural network.

Follow-up and monitoring. As with other NN financial applications, the user must constantly monitor how well the neural networks are performing in predicting claims payouts and securities behavior and have a plan for prompt retraining when certain conditions are met.

Commercial software packages. Another useful product from HNC Software, Inc., is *MIRA®,* (Micro Insurance Reserve Analysis), a neural network-based system originally designed for workers compensation claim reserves. *ProviderCompare®,* also from HNC, is a physician-profiling system to help establish medical claim reserves. HNC integrates these systems with its *Predictive Software Solutions* to estimate reserve requirements.

Many large financial institutions use an advanced NN package from the AND Corporation to forecast the performance of derivatives and commodities and for currency trading, among other financial market predictions. The package also includes an automated actuarial system.

Stock 100, Inc., a financial engineering consulting firm, uses proprietary NN software to advise financial institutions. In conjunction with CyBerCorp, Inc., Stock 100 offers on-line trading as well as continuous real-time prediction of securities performance.

Trading Advisory Services

Financial advisors give customers investment advice about specific securities and about what's happening in various market segments. These advisors offer tips on the timing of buying and selling securities that respond to criteria established by the customer (e.g., short-term or long-term position, high-risk or low-risk position). The success of a trading advisory firm depends on how well it provides the right signals to customers, whether in real time (continuously) or on a periodic basis (daily, weekly, or monthly reports).

Trading Advice

With a real-time data feed advisory service, throughout the trading day the advisor issues recommendations to customers about securities the customers already hold and about securities the advisor is recommending. Because the execution time of a neural network is so rapid, the prediction model can operate in a fraction of a second, making adjustments to recommendations made just a few minutes before. It can predict the price of a security in 30 minutes, an hour, or on some future date on the basis of the current price and most recent history. One researcher found that using a neural network he could accurately predict the price of gold futures 30 minutes from the current quote 92 percent of the time. Some day traders and portfolio managers use real-time feeds both to minimize losses and to take early advantage of what will be a major swing in the price of a security.

Planning the NN application. If the trading advice deals with specific securities or with specific markets or market segments, the software may require a series of NNs, one for each security or segment. The advisor may need some up-front processing consolidate market or market segment data before running the NN series. Finally, the NN must be linked to software to communicate the results of the NN prediction to customers. Most financial advisors embed the NNs in their on-line or e-commerce services to customers.

Identifying and selecting data. Data sources for a trading advisor NN application vary depending on the advisor but generally combine advisor internal databases, databases from subscription services, and direct feed from the exchanges where the securities of interest are traded. Data selected, as with other securities-related applications, includes security, market segment, market data from past experience of the advisor, and outcomes needed to accomplish objectives.

Training the NN to provide trading advice. Results of past securities assessments, internal or external, are the best means of training a neural network to assess individual securities in the future. Otherwise, a neural network used for trading advice is trained just like any other NN (see pp. 63).

Using the NN to provide trading advice. The trained network can be integrated with other decision support systems to monitor movements in capital markets and then advise customers directly or through the trading advisor to buy, sell, or hold the securities monitored. By monitoring market changes to predict upswings and downswings in the market in the market place, the NN can signal buy and sell recommendations to both the customer and the advisor instantaneously.

Retraining the NN. As the advisor gains more experience with securities markets, adding new experience to the training set can retrain the neural network.

Follow-up and monitoring. As with other NN applications, the user should have in place a plan for retraining or other follow-up if regular monitoring reveals that the NN is not predicting the behavior of securities as well as is necessary.

Commercial software packages and services. DynaMind® offers NN-based software that does time series analyses of trading in stocks and commodities.

Lester Ingber Research, of Chicago, IL, provides NN-based consulting services to companies trading commodities futures.

Trading Rules and Rule Generators

Service financial advisors provide not only trading rules but also trading rule generators. The ability of a neural network to recognize patterns in complex data enables the advisor to formulate trading rules based on recognizable data patterns associated with the trading objectives. NNs can also be embedded in rule-generator software that monitors changes in securities and markets and adjusts the trading rules to suit the environment. The speed and effectiveness of NNs makes them a powerful platform in the trading rules area.

Planning the NN application. If the trading rules or generator deal with specific securities or specific market segments, the software may require series of neural networks, one for each. The advisor may need some up-front processing to accumulate and consolidate market data before running the NN series. The NN must also be linked to software that communicates the NN results promptly to customers. Most financial advisors embed the NNs in their on-line or e-commerce services.

Identifying and selecting data. As with similar NN functions, the data sources generally involve a combination of internal and subscription databases along with direct feeds from the exchanges where securities of interest are traded.

Training the NN for trading rules and rule generators. Again, results of past securities assessments, internal or external, are the best means of training a neural network to assess individual securities in the future. Otherwise, a neural network used for reserving functions is trained just like any other NN (see pp. 63).

Using the NN to formulate trading rules and rule generators. The trained network can be integrated with other decision support systems to monitor movements in capital markets and to set rules for buying, selling, or holding decisions about securities the customer holds or the advisor tracks. By monitoring changes in the marketplace, the neural network can make predictions about market direction that can be used not only to signal buy and sell decisions but also to signal the success or failure of a trading rule instantaneously.

Retraining the NN. Once again, as the situation changes, the trading rules must change. As the advisor gains more experience with outcomes of trading rules, adding new experiences to the training set can retrain the neural network.

Follow-up and monitoring. As with any other NN application, performance must be monitored constantly. The user should have in place procedures to follow up whenever NN performance needs to be adjusted.

Commercial software packages and services. Scientific Consultant Services offers *Trading Simmulator.*

Portfolio Management

Portfolio managers generally have a fiduciary responsibility to handle assets according to the portfolio charter or other agreements. They are thus the investors for their portfolios and need to monitor what is happening in the securities' markets to seek out optimum returns for the assets with which they are entrusted.

The three areas of portfolio management in which NNs provide significant help are (1) securities selection, (2) consistent application of constraints, and (3) hedge advisories.

The goals and objectives of a portfolio will narrow the choice of securities for investment. The manager identifies the general types of securities that conform to the portfolio charter or agreement, and then a neural network can be used to select those that optimize the return on the portfolio. Assessment of (a) individual securities and (b) segments of securities markets are the two main areas where portfolio managers use NNs to fulfill their fiduciary responsibilities.

Planning the NN application. Once again there are minor differences between the two areas the NN is asked to assess, but the execution of the NN doesn't change. To assess a given investment opportunity, NN design begins with identification of security attributes that relate directly to the objectives of the portfolio agreement. The manager may use indirect measures of achieving the investment objective.

Identifying and selecting data. Data sources that support portfolio goals will generally be a combination of internal and external databases, such as those from subscription services. Data selected includes security, market segment, market data from past experience of the manager or the outside source, and outcomes that will accomplish portfolio objectives.

Training the NN for portfolio management. The best way to train a portfolio management NN is with the results of past assessments of securities. Otherwise, a neural network used for reserving functions is trained just like any other NN (see pp. 63). Don't forget the holdout sample.

Using the NN for portfolio management. The trained network is integrated with other decision support systems to monitor movements in capital markets and to initiate instantaneous buy, sell, or hold decisions.

Retraining the NN. As the portfolio manager gains more experience with the markets, and as those markets change, adding new experience to the training set can retrain the neural network.

Follow-up and monitoring. It is important to have in place a procedure for following up immediately when NN performance begins to deteriorate.

Commercial software packages and services. DynaMind® offers NN-based software for time series analyses of stock and commodity trading.

Applied Analytic Systems, Inc., of Pittsburgh is a consulting service that helps companies develop proprietary NN-based financial market prediction software.

Real Estate Brokers

Real estate brokers help homeowners and homebuyers; they place homes on the market and negotiate the sale of homes. NNs can help them set the right listing price for homes and prequalify the financial stability of potential buyers or developers. Both of these areas depend on complex data that NNs can massage to significantly improve broker services.

In helping homeowners set the listing or asking price, real estate brokers look at current conditions in the local market, the condition of the properties, the amenities the homes offer, their size, and the neighborhood. In most local markets, the real estate association publishes a listing (the Multiple Listing Service—MLS) of all properties currently on the market with asking prices and description of the properties so that other brokers are aware of what properties are available, thus increasing the number of potential buyers. The MLS also contains data, by neighborhood, on recent sales or other outcomes for properties that have been listed. Brokers use NNs to analyze this database as they set asking prices for newly offered properties.

As for prequalifying, brokers deal with two types of buyers, those making an offer on a property the broker listed and those using the broker to find a property. Brokers also deal with developers, listing homes they have built or refurbished. The financial stability of a potential buyer or a developer directly affects the ability of the broker to close a transaction. Buyers usually need financing to close a transaction. Developers must be able to survive financially for as long as it takes to sell their properties.

Predicting the Selling Price

There are many formulas for establishing an asking price for a property; the challenge is to find the price that optimizes the value of the property. Some brokers use NNs to predict, with significant accuracy, the eventual selling price. The listing price can then be established to accomplish a quick sale or a maximum price sale or devise whatever other strategy is needed to meet the owner's objectives. The charge to the broker may be, for instance, to obtain the highest possible price for a property or to sell the property as quickly as possible for whatever price is reasonable. This prediction is also a benchmark against which to compare offers received from potential buyers.

Planning the NN application. The design of the NN must incorporate all the attributes that relate directly to the selling price of the offered property. As a fallback, the NN generally uses the selling price of recently sold similar properties as the desired objective.

Identifying and selecting data. The data sources used for real estate NN applications vary depending on the broker's criteria but generally combine internal databases and the MLS. Data selected include condition of sold properties, their amenities, the size of the home, and the neighborhood. Including data from adjacent neighborhoods enhances the NN's ability to more accurately predict the selling price of a home. The broker should review other subscription databases for potential inclusion in the input data set.

Training the NN to predict selling price. A neural network used to predict the selling price of a property is trained just like any other NN (see pp. 63). In this case the massive set would be all sales of properties in the MLS.

Using the NN in listing properties. A trained NN can be integrated with other decision support systems to establish the listing price of a property and monitor offers made on it.

Retraining the NN. As more information becomes available about listings and transactions in given neighborhoods, it can be added to the input data to retrain the neural network.

Follow-up and monitoring. Given that property markets can change fairly quickly, brokers using NNs should be constantly monitoring them and the environment to make sure the NN is performing well in predicting the behavior of selling prices, and should have procedures in place to retrain the NN promptly whenever the need arises.

Commercial software packages and services. Transamerica Intellitech, Inc. of Sacramento, CA, offers *AREAS,* a neural network-based software package originally developed by HNC Software, Inc. to predict the market price of real estate. Users include all types of real estate professionals, among them title companies, lenders, agents, and appraisers.

Prequalifying Buyers and Developers

In doing prequalifying, real estate brokers have two concerns (1) the financial stability of the potential buyer or developer, and (2) the potential for fraud. The goal of using NNs is to reduce the time and cost of processing real estate loan applications and to improve the quality of offers so as to minimize delay in closing or avoid lost sales. Many brokers use NNs successfully for on-line prescreening of potential buyers to determine whether to decline the offer or recommend that the seller make a counter-offer. Some brokers use NNs to make sure developers will survive long enough for listed properties to be sold at a reasonable price. Agents at branch real estate offices can financially qualify buyers using a trained NN that is part of the general operating software.

Planning the NN. As usual, the design of a neural network begins with a clear and precise definition of what the network should accomplish, when it should do it, and how the results should be communicated. For prescreening the financial stability of potential buyers, the decisions about data, the type of NN, and the outcome can be stated succinctly. When an applicant submits a financial position form to the broker, the data is entered into a computer, which runs the NN. The NN makes a preliminary determination about the creditworthiness of the applicant. While it does not assure that a financial institution will underwrite a mortgage, prescreening helps the broker avoid costly delays and voided closings when the buyer cannot obtain financing.

Where developers are concerned, the broker uses a financial statement application and other data such as credit bureau reports and records on the particular developer's history. The goal in reviewing the data is to determine financial stability of the developer during the listing period. NNs can accurately identify many of the relationships that exist in large bodies of developer data.

Identifying and selecting data. Financial disclosure data varies significantly from a few lines on a standard "offer to buy" form to a complete statement of net worth with attached supplementary financial statements. In general, the more data, the better the NN can decide, though sometimes a more limited set of data can also yield good results. Generally speaking, for the NN the broker needs data from both successful and failed credit experiences. The network learns when it is shown both successful financial positions and those applications that were accepted in the past but have failed. Without data on the unsuccessful offers, it will be impossible to train a neural network. If possible, it is best to run several pilot projects of a small sample (100 to 150) of offers; by varying the number of inputs to the network, the broker can identify the best combinations of inputs for an effective NN. In general, there should be as much data as possible about the developer and about economic conditions from at least the last two quarters to preferably the last five years.

Economic conditions can affect the ability of the potential buyer to obtain mortgage financing or the developer to maintain solvency.

This is another case where data preparation of the data is hard work, because NNs require numerical data representation. Codes that use alphabetical characters must be converted, and there should be large enough gaps between one number and the next to allow for appropriate discrimination between observations.

Training the NN network application. A NN used to predict financial stability of real estate buyers and developers is trained just like any other NN (see pp. 63).

Using the NN. Once trained, the NN can be installed as a subroutine within whatever operational software the broker uses to process offers or make other decisions. If the NN is executed at the end of the data entry program or just after all the validity check routines have been executed, the potential buyer can be notified immediately about whether or not the application is accepted.

In most situations, the neural network makes a recommendation to the broker, who uses professional judgment to make the final determination. But sometimes the network itself makes the final call.

Retraining the NN. Whenever there is a change in the economic environment or general profiles of financial stability, the broker can add the new experience to the training set to retrain the neural network.

Follow-up and monitoring. Whenever accepted transactions fail to close, brokers using NNs should figure out why the NN failed to detect noncreditworthy applicants and use predetermined follow-up procedures to bring the NN back into alignment.

Commercial software packages. MGIC Investor Services Corporation, an affiliate of the Mortgage Guaranty Insurance Corporation, offers the NN-based *Loan Performance Score,* a software package that predicts the likelihood that a mortgage will go into foreclosure within four years.

Flagging the Potential for Fraud

Developers often attempt to stay solvent or avoid bankruptcy proceedings by selling properties prematurely. Properties a broker has listed may not close because the developer cannot complete construction or because unpaid subcontractors and others have placed liens on the title of the property.

False financial reports are the most common form of developer fraud against brokers. A neural network can flag potentially fraudulent financial reports so that the broker knows that more investigation is necessary before accepting properties for listing.

Planning the NN application. For this type of application, the NN should be designed to recognize patterns of data in financial statements or disclosures that indicate possible fraud. Though financially stressed developers may provide data that seems convincing to the real estate broker, fictitious reports often betray subtle patterns to a neural network.

Identifying and selecting data. Observations useful in this type of NN application may include the developer's previous financial disclosures, past encounters with the developer, number of properties for sale, type of properties, characteristics of the developer (from the original application), and other account data. The real estate broker should use data from several accounts that have been subjected to fraud. The more data the better for this type of neural network—two to ten years of history is not too much.

Training the NN. A neural network used to predict financial stability of real estate buyers and developers is trained just like any other NN. (see pp. 63). The massive set here would be all experience with all developers; the simplified set would be a representative sample of legitimate and fraudulent financial disclosures from developers. In any case, part of the training includes a holdout sample to test the trained network. This test provides assurance that the NN can accurately distinguish between legitimate financial disclosures and those that may involve fraud.

Using the NN to detect fraud by developers. A trained NN can be installed as a subroutine within any of the operational software a real estate broker uses to make decisions about the financial stability of developers who want to list properties through the broker. The NN is embedded in the software near the end of processing the application materials, after all inputs have been obtained. Usually, the NN simply generates an alert to the broker to do more investigation.

Retraining the NN. Over time circumstances change and the patterns of fraudulent financial disclosures by developers also change. As the broker gains more experience with more developers and with how they make fraudulent financial disclosures, the NN can be retrained by adding these new experiences to the training set. Training on a monthly or quarterly basis is most effective.

Follow-up and monitoring. Brokers need to keep an eye on how well the NN is predicting developer performance and follow up at once whenever they see a problem of underperformance.

Commercial software packages and services. HNC Software, Inc. offers *Eagle™*, a neural network-based merchant risk management system designed to protect acquirers from emerging fraud trends.

Investment Banking

Investment banks underwrite the issuance of securities. As part of its services, the investment bank performs due diligence on the claims made by the issuer of the securities in the prospectus filed with the Securities and Exchange Commission (SEC). If all is well, the shares are offered for sale and issued to the public. NNs provide significant value to investment banks in two areas of due diligence (1) assessing the financial position of the company and (2) identifying fraudulent financial reporting.

NNs can help an investment bank determinate the risks associated with a given security. The investment banker must determine the interest rate for bonds or the selling price for stock issues of customers. Assessing factors like how the industry as well as the issuer is performing and what are the economic conditions in the market for this particular security provides data that affects the offering price. The investment bank can use NNs to predict how the market will react to particular securities. It can also use NNs to flag the potential for fraudulent financial reporting by the issuer that could affect the offering price and could also lead to liability for the investment bank.

Setting the Selling Price for Securities

The investment bank underwriting a securities issue hopes to sell all the securities on the day of issue or shortly thereafter at the offered price. The longer the investment bank holds the securities, the less likely it will recoup its investment in those securities. Accurately predicting a selling price at which the market will buy most, if not all, of an issue within a few days can make the difference between success and only a marginal return. NNs can do this.

Planning the NN application. Effective NN design requires the investment bank to identify attributes that relate directly to the selling price of securities.

Identifying and selecting data. Data selected for this type of NN include the securities offered, other securities of the offering entity, economic conditions, market segment, market data from past experience at the investment bank, and success in selling similar securities. The manager should review outside subscription services for potential inclusion in the input data set.

Training the NN to predict selling price. To train a neural network to assess individual securities, results of past assessments are most useful. Otherwise, an investment banking NN is trained just like any other NN (see pp. 63).

Using the NN to predict the optimal selling price of a security. A trained investment banking NN can be integrated with other decision support systems to establish the offering interest rate of credit instruments or selling price of stock issues.

Retraining the NN. Patterns for new security offerings and activity in different securities markets change often. As the investment bank gains more experience with securities markets and new offerings, the NN can be retrained with the new experience.

Follow-up and monitoring. Investment banks using NNs to help price securities should be constantly watching to see how well the NNs are predicting selling prices of new securities. If performance is not satisfactory, there should be follow-up procedures in place to bring performance into alignment with goals.

Commercial software packages. Watchdog Investment Monitoring System is popular for investment analysis and management. It incorporates financial statement analysis and financial forecasting.

Flagging Potential Fraud

Company managers, officers, and significant investors as well as other individuals often attempt to influence the issue price. When that activity is directed toward fraud in the initial offering of securities, it usually starts with falsified financial disclosures designed to increase the selling price. Most fraudulent financial disclosures have a pattern of manipulation that traditional methods may fail to recognize. NNs do recognize these patterns. A neural network, by flagging potentially fraudulent financial disclosures, can alert the investment bank to the need for more intensive due diligence.

Planning the NN application. For this application, the NN should be designed to recognize data patterns associated with fraudulent financial disclosures. The perpetrator of the fraud will impart financial data that shows higher than actual earnings, potential earnings, cash flow, potential cash flow, and liquidity and other indicators that investors and analysts monitor. To accomplish this window dressing, the perpetrator of the fraud takes into account the typical models investors and analysts rely on in assessing the indicator of interest. Consequently, the fraud is targeted to one or a few particular items of disclosure, often ignoring the others. This causes an irregular pattern in the financial data.

Investment banks use NNs to alert due diligence teams to the possibility that the financial data is being manipulated.

Identifying and selecting data. Observations that are incorporated into the NN include past financial disclosures for a given client as well as other financial data. The investment bank should use data from clients whose submitted disclosures have proved fraudulent or even as well as those financial disclosures that are free of fraud. For this application, the history should go back at least five years and even as far back as twenty years.

Training the NN. Investment banks train their fraud detection NNs like any other NN is trained (see pp. 63). A massive set here would be all experience of financial disclosures for all clients and a simplified set would be a representative sample of legitimate and fraudulent financial disclosures from clients. A holdout test can provide assurance that the NN can accurately distinguish between legitimate financial disclosures and those that may involve fraud.

Using the NN for fraud detection. The trained network can be installed as a subroutine within any of the operational software the investment bank uses for due diligence on an issuing company. It is embedded in the software near the end of processing of the initial data, after all inputs have been obtained but before any site visits, so that the visitation team can be alerted to potential problems. When fraud is suspected, the visitation team doing due diligence fieldwork can decide beforehand what additional evidence to gather to discover the existence and extent of any fraud.

Retraining the NN. As the patterns of fraudulent financial disclosures associated with new securities' offerings become clearer or change, the new experience can be used to retrain the NN.

Follow-up and monitoring. NNs must be monitored to make sure they are actually detecting fraudulent financial disclosures. If not, there should be follow-up procedures in place to bring performance up to par.

Commercial software packages and services. *Eagle* is a neural network-based package from HNC Software, Inc. that is a merchant risk management system designed to protect acquirers from emerging fraud trends.

Retail Securities Brokers

Retail brokerage firms provide a wide range of services, many of which have already been addressed in the sections on trading advisors and portfolio management, but an area of special concern for them is managing margin accounts. Small swings in market prices that have little impact on most investors can be significant for customers investing on margin. These small swings can be either a boon or disaster, depending on the direction of the change and the width of the margin.

NNs can be used to track short-term movements of specific securities online. They can give the broker early warnings or predictions on the likely movement of the price of a security. The NN can then signal when a margin account is stressed so the broker can notify the customer to take action before the margin evaporates. Providing this service can reduce margin calls and improve customer goodwill.

Planning the NN application. Design of a neural network for assessing short-term price changes of a given security requires identifying attributes that relate directly to predicting small movements in securities prices. Specifically, the broker must determine the degree of price change and the time frame that is critical. If the minimum margin is 20 percent rather than 5 percent or 10 percent, more price fluctuation is allowable before margin calls. Timing is determined by the average time it takes to notify a customer and obtain a sell order. Identifying a suitable way to meet the objective is crucial to successful use of a neural network to manage margin accounts.

Identifying and selecting data. Data selected for this application include security, market segment, market data from the broker's past experience, and outcomes desired for accomplishing objectives. The broker should review subscription services and other databases for potential inclusion in the input data set.

Training the NN for margin account management. Results of past assessments are highly effective in training a neural network to predict short-term price movements for individual securities. Otherwise, these NNs are trained just like any other (see pp. 63). The massive set is all margin account experience; the simplified set is a representative sample of margin account experience including called, salvaged, and good accounts.

Using the NN to manage margin accounts. The trained network can be integrated with other decision support systems to monitor, on-line, price movements of specific securities for margin account customers, flagging those accounts that need immediate warning and action.

Retraining the NN. As circumstances and the patterns of short-term price movements change and as the broker gains more experience with both margin management and the securities markets, the NN can be retrained by adding new experience to the training set.

Follow-up and monitoring. Retail securities brokers using NNs to manage margin accounts should be constantly checking to see how good the NNs are at predicting short-term security price behavior. Follow-up procedures should be in place to bring straying NNs back on target.

Commercial software packages. HNC Software, Inc. offers *ProfitMax® Margin Manager,* which tells the margin manager the probability that a given account will need to go on margin call during a specified period and the potential loss to the account if there is a call.

CHAPTER 5

Artificial Intelligence in Accounting

The two main applications of AI in accounting so far are expert system and neural network technologies. While fuzzy set, genetic algorithms, and chaotic models are currently being considered for their potential application to accounting, ESs and NNs are already working well in a number of accounting areas.

EXPERT SYSTEM APPLICATIONS

ESs are useful in accounting because financial and managerial accounting, income taxes, and auditing are all rule-based disciplines. Moreover, most accounting applications are already computerized. To the extent that computerized accounting systems follow structured programming techniques, converting them into a knowledge-based system with an inference engine requires only a knowledge engineer and an application expert.

The areas of accounting where ESs are most commonly applied are cash management, marketable securities, accounts receivable, prepaid expenses, inventory, property, plant and equipment, investments, accounts payable, payroll, long-term debt, contingent liabilities, deferred income tax, earnings per share, auditing, and income taxes.

Cash Management

All accounting systems that use a bank account for most cash transactions are confronted with the monthly bank reconciliation. The steps associated with bank reconciliation are ideally suited to expert systems. The ES queries the reconciliation clerk to identify outstanding checks, deposits in transit, special collections, bank charges, and insufficient funds checks deposited. The system then makes recommendations on what to do about these items to balance the bank statement with the book balance and to deal with problems like duplicate posting of checks, coding errors on checks, and forgeries.

It makes sense to use an ES for bank account reconciliation because the rules for reconciliation of a bank account are stable; they do not change over time. Once the system is in place, it requires little or no maintenance. A second advantage of using an ES derives from the infrequency of bank reconciliations. Because bank statements are issued monthly, for most companies reconciliation is done only once a month; so most clerks must review the procedures for bank reconciliation before or during the process. This can result in delays or inefficiency in preparing the reconciliation. Using an ES is a highly cost-effective use of AI in accounting.

Securities

Expert systems can be used as an internal control to monitor selection and performance of securities that are bought to put idle cash to work. ES can execute investments, sell orders, and hold orders based on the original authorization from the board of directors and the policies management has established, which are incorporated into the ES knowledge base and can be queried by the inference engine monitoring market conditions throughout the day.

Accounts Receivable

Activities associated with accounts receivable that are appropriate for ESs are order validation, credit approval, and invoicing and updating subsidiary ledgers. These activities depend on the type of complex decisions that can be supported by an ES.

To validate an order, it is necessary to make sure the customer exists, that the purchase order is valid and signed by an appropriate representative of the customer, that the customer is authorized to receive goods ordered, that the goods ordered can be delivered by the time specified in the order, and that the sales representative is noted (especially if payment is on commission). All these activities have permutations such as exceptions considerations that make the decisions complex; an ES can enhance the quality of those decisions. It can expedite routine sales orders, and it can speed processing of even nonroutine orders within certain limits. The rare cases needing management review can be flagged and immediately brought to the attention of the sales manager.

Granting credit involves deciding whether to grant credit to a new customer and determining if a current customer is over the approved limit either in general or for a given order. Making the initial credit decision typically means that the potential customer must provide credit reports, audited financial statements, and other important disclosures about owners and management if it is a business. This information is reviewed and sometimes entered into a credit scoring system. Often the credit scoring system uses an ES to make the credit decision and set the credit limits. Rules for granting credit under different conditions can be installed in the ES knowledge base, and credit staff can invoke the inference engine as they process the application form and review the information submitted. The inference engine can be designed to mirror the credit application form.

To approve credit for an existing customer it is necessary to see if the total of the current purchase added to the customer's current balance will exceed the customer's credit limit. While this process appears simplistic, the process of deciding to reject the order can become very complex. A series of decision rules coded into an ES knowledge base makes it possible to process many routine orders where the credit limit can be extended (limit was established years ago and never revised, for instance, or the account is current and the excess over the limit is within 10 percent of the total ordered). These rules can be exhausted relatively quickly. If the order is still unacceptable, it can be flagged for management review.

Invoicing involves checking the original sales order with credit approvals and shipping records, then extending the unit prices and quantities shipped and adding shipping charges and sales taxes. The decision rules associated with it are very adaptable to expert systems. A knowledge base and the inference engine can be tailored to the specific company. Use of expert systems reduces the need for highly skilled accounts receivable clerks to do invoicing. An ES can automatically post the details to the customer's subsidiary ledger record by generating the sales transaction journal and immediately posting to the customer's account.

Sales returns and allowances require complex steps that are somewhat out of the ordinary, but an ES can readily apply the decision rules associated with determining if returned or damaged goods have been received and the amount of credit to be granted the customer. The credit memorandum entry can be automatically generated, entered into the appropriate journal, and posted to the subsidiary ledger immediately.

Lastly, ESs are very effective at processing invoice payments by reconciling the payments and any credit memos associated with sales returns or allowances. The decision rules about which invoice to apply the payment to, whether or not there are any credit memos, and whether there is any outstanding balance associated with a given invoice can be easily programmed into the knowledge base, and the inference engine can be adapted to facilitate collections.

Prepaid Expenses

Prepaid expenses generally expire over time or on the occurrence of a specific event. An ES can efficiently monitor the passage of time and check to see whether the event has occurred to determine how much prepayment has expired for a given item. The knowledge base can be established with the rules for expiration and the inference engine set up to be executed at the close of the accounting cycle for time-based expiration rules. For event-based expirations, the ES can be embedded in the transactions processing program.

Inventory

The two classes of inventory are merchandise and manufacturing inventories. Both classes are subject to two rules of valuation, the cost flow assumption rule (first-in, first-out [FIFO], last-in, first-out [LIFO], weighted average, and specific identification) and the "lower of cost or market" rule. ESs handle both rules extremely well. Each type of inventory also has unique and complex decisions that ES also handle easily. The next section discusses expert systems that deal with the valuation issues both types of inventory have in common and a later section discusses how ESs can be used in manufacturing inventory.

The cost flow valuation is a function of whether or not the assumption is FIFO, LIFO, weighted average, or specific identification. Once that decision is made and the cost of the ending inventory found, the lower of cost or market rule can be applied. This rule compares the inventory cost with the market value at the balance sheet date. These decisions can easily be accommodated using an expert system with the decision rules embedded in the knowledge base and the inference engine embedded in the inventory pricing software associated with the taking of physical inventory or year-end closing procedures.

The rules for determining the appropriate price for each inventory item are coded in the knowledge base. The inference engine accesses the inventory database, searches the knowledge base for the rules associated with the cost flow assumption, and sets the price to associate with the ending inventory count. The inference engine then accesses the current inventory market or selling price database for the current market or selling price, using the knowledge base to determine the price associated with the item. The ES then updates the inventory database. If the current market or selling price is less than the cost-based price determined under FIFO, LIFO, or weighted average, the ES updates the inventory database with the lower market value.

The ES then generates a series of adjusting entries to update the general ledger database for preparing financial statements and a schedule of ending inventory prices at cost and those items where current market or selling price is lower than cost.

Manufacturing Conversion of Inventory—Cost Accounting

ES provide significant value-added for inventory valuation in cost-based accounting systems, particularly activity-based costing (ABC) and standard costing systems. In these systems, costs are labeled either direct or indirect. Direct costs are those that can be directly identified and measured against units of product produced. These costs typically include labor, materials, and power, and in some cases depreciation. Indirect costs either cannot be identified or cannot be measured or associated directly with products produced. These costs, referred to as overhead, are accumulated into cost pools. At the end of the accounting period overhead costs are allocated to work-in-process (WIP) and finished goods (FG) inventory using allocation rules set by management. Allocating overhead can be very complex. Because it's rule-based, however, it's an ideal application for expert systems.

The traditional method is to allocate overhead to products based on materials cost, materials usage, labor cost, labor usage, machine hours used, power consumed, or some combination of those measures. Occasionally other measures are used, such as number of employees involved or square footage of factory space occupied during production of given goods. Some systems capture costs of conversion of the inventory using job order, process, or some hybrid method of measurement. In all cases, the basis for traditional allocation derives from the relative amount of direct costs used in the inventory conversion process.

The traditional rules for allocating overhead costs are easily coded into an ES knowledge base. The inference engine for determining the appropriate rule and amount of allocation is embedded in the cost accounting system for job costing or for period-end processing of cost information. The advantage of using ESs to perform these allocations is the ease of setting complex allocation rules and the minimization of errors in applying the rules.

ABC Allocation Methods

ABC costing has a consumption function and a complex measurement function that are ideal for ES applications. ABC costing is done in three stages. The first stage is to measure activities of each support organizational unit and its costs. The second stage is to accumulate activity costs across all organizational units. The third stage is to absorb the activity costs associated with the product based on consumption of that activity by the product. This three-stage process complicates costing significantly. Nevertheless, most of the measurement of activities is computerized, as is the accumulation across organizational units. The rules for absorption of the activity cost pools lend themselves to coding in an ES knowledge base. The inference engine is embedded in the cost accounting system. Using information contained in the inventory databases (WIP and FG inventories), the inference engine searches the knowledge base to determine the appropriate rules for absorption of the cost pool. The inference engine then determines the overhead cost to assign to the FG and WIP inventory items and generates transactions to update the inventory ledger and general ledger.

Standard Cost Accounting Systems

Standard cost accounting systems establish and assign inventory (WIP and FG) costs using a unit cost standard. These systems are heavily used for homogeneous processing systems and homogeneous products or product lines. These systems have three phases of activities (1) standard cost build-up, (2) standard cost assignment, and (3) standard cost reconciliation.

Standard cost build-up consists of determining the individual direct cost elements and a reasonable estimate of the amount of overhead needed to produce one unit of a given product. This phase occurs before the accounting period and typically is the basis for preparing the detailed production budget for the coming year. In the most simplistic case, next year's standard is calculated based on an average of the current year's actual costs per unit of product but taking into account anticipated changes in materials, labor, equipment, and production procedures as well as anticipated price changes for each item of production. The effect of each of these elements on the individual cost elements of each product produced constitutes a decision rule. ESs assist in the standard cost build-up process by capturing the rules developed from estimates in the knowledge base; they search for the appropriate rule for each cost element associated with each product to set the standard and they update the standard cost database for the next year.

Once the standards are established, standard cost systems apply, starting with setting the details of the production budget before the beginning of the year. Once the production budget is in place, standard costs are applied to products during the production process. Periodically (usually monthly), the standard cost system compares actual costs with total costs applied to existing production using the unit standards. The outcome is a standard cost performance report. Generally, these reports generate variances of actual costs from standard costs. Because these reports tend to be limited, it takes significant interpretation for managers to recognize issues and take action. ES can be very valuable in interpreting variances and making recommendations for action.

Interpreting variances follows rules that become complex very quickly and requires intimate knowledge of the production process to identify problems and to determine a course of action. The rules for how to interpret specific variances can be coded into the ES knowledge base; the inference engine searches the knowledge base along with the inventory database to identify the most likely cause of the variance. Through the user interface the ES makes recommendations to the manager responsible for potential courses of actions and can even rank or calculate the probability for each potential cause of the variance. For example, a labor rate variance for a given department in a process costing system may arise from use of the wrong type of labor (more skilled or less skilled) than management planned in setting the standard cost. Alternatively, the variance may arise because of a significant deviation in volume of activity from the average volume used in establishing the standard. It could arise from the use of more expensive or less expensive materials, or use of new technology or equipment.

This simplistic example demonstrates how quickly the potentials become complex. An ES can examine a number of potential causes and indicate which is most likely or produce a ranking list of likely causes of the variance. This can save significant management time, resulting in more efficient production.

Variances in the cost system result in over or underapplied costs. These are charged to overhead accounts throughout the accounting period. At the end of the year, any balances in the over or underapplied accounts are closed (proportionately) to inventory and cost of goods sold, though theoretically, the year-end balances in these should be at or near zero. In some months there are balances due to cyclical fluctuations in production; an overapplied amount in one month should be offset by an underapplied amount in another month. Nevertheless, the standards are set based on estimates of what will occur in future periods and often there are significant balances in these accounts at year-end that must be apportioned to WIP and FG on some reasonable basis. Expert systems assist in the reconciliation and application of year-end variance balances to WIP and FG. The rules for dealing with the variance accounts can be coded into the knowledge base. The inference engine then determines how an expert would apply the variance balances between the WIP and FG and the ES makes the recommendations through the user interfaces.

There are several important ES packages that offer advice for capital and operations planning and budgeting applications. Arthur D. Little's Expert System performs variance analysis and explains the reasons for significant deviations. *Financial Advisor*, developed at MIT's Sloan School of Management, appraises capital investments in property, plant, and equipment. It also gives advice on projects and products. Azure Software's *Project Analysis for Capital Equipment Expenditures* analyzes and reports on the feasibility of acquiring fixed assets. Cash Value is a commercial expert system for capital projects planning.

Accounts Payable

Accounts payable processing covers the purchasing and payment for goods and services used by the company for operations. Relevant activities are the purchase request, choice of vendor, ordering goods or services, receiving the goods or services, and paying for the goods or services.

Requesting goods or services is usually initiated with a purchase requisition issued by an operating unit when goods are needed. When significant lead times are needed, decision rules typically guide managers in determining when to make the request. The objective is to minimize inventory-holding costs while still be taking advantage of potential quantity discounts, all without negatively affecting operations. These complex rules are encoded in the knowledge base and after the inference engine is invoked, the user interface recommends which goods and services to request; it may even produce the actual purchase requisition.

The next step is purchasing or procurement. This step uses another set of rules, some for internal control and some for efficient and effective operation of the company—rules for obtaining bids, verification of the reputation of vendors, assessing past performance, and evaluating the price competitiveness and quality of goods or services provided. When goods or services are being ordered, the inference engine queries the knowledge base where the rules are coded to find the optimal vendor.

The next stage is the receipt of the goods or services. Some companies require very complex goods or services, and the rules are not only themselves complex but they often overlap. When receipt must be verified, the receiving department invokes the inference engine to search the rules. The resulting recommendation incorporates not only verification of the quantity of goods and services received, but also an evaluation of how closely the goods or services match the technical specifications of the original purchase order, along with suggestions for purchase returns and allowances when there is an incomplete match.

The final step in the process is payment for the goods or services. Internal control requires matching the purchase order, a receipt acknowledgement, an invoice from the vendor, and any adjustments for purchase returns and allowances. Once this is done, a voucher is produced for payment. Good cash management stipulates that vouchers should be paid at or near the due date on the vendor invoice, and discounts should be taken when available. Rules for voucher processing are coded into the ES knowledge base. The inference engine queries the purchase order database, the receiving database for receipt acknowledgements, and the accounts payable database for invoices and debit memoranda. Once all is verified the ES can schedule the appropriate payment date for the voucher through the user interface.

Payroll

Payroll involves many activities, from hiring, promotion, demotion, and termination of employees to tracking the basis for pay and other employee benefits and calculating and paying employees. Complex rules on status changes, policies, and procedures are coded into the ES knowledge base. When an employee is hired, promoted, demoted, terminated, or retired, the inference engine searches the human resources database and queries the manager about the employee action. When the queries are answered, the inference engine makes a recommendation about further actions to be taken with respect to the employee. In addition, the ES generates payroll system status change transactions to update the payroll master files.

Long-Term Debt

Long-term debt includes notes, mortgages, bonds, capital leases, and certain contingent liabilities, all of which have very complex accounting rules that lend themselves well to expert systems. The accounting rules for issuance, retirement, defeasance, reissue, conversion to another instrument, interest calculations, amortization of any premium or discount, and economic impact for footnote disclosure call can be encoded into the ES knowledge base. At the end of the accounting period, the inference engine queries the knowledge base and the accountant to determine the appropriate adjusting entries. The ES recommends to the accountant through the user interface which entries to make as well as footnote disclosures for events that have occurred. Some ESs generate the adjusting entries for automatic update of the general ledger.

Deferred Income Taxes

Deferred income taxes arise from the temporary differences that result from use of different accounting principles for measuring and recording transactions for financial and income tax reporting of economic activities. ESs provide significant help in determining the effects of these differences. Given the number of potential differences in accounting principles between financial reporting and tax reporting, most companies are likely to have temporary differences between the tax expense reported on the financial statements and the tax payment made with reported tax statements. ESs encode the potential differences in the knowledge base. At year-end, the accountant invokes the inference engine, which ascertains which accounting principles were used for financial and which for tax purposes. The inference engine then calculates the difference for each account and through the user interface communicates the adjusting entry to recognize the deferred taxes for the year. Some ESs also automatically generate adjusting entries to update the general ledger system.

Earnings per Share

Earnings per share calculations are complex because they must take into account the dilutive effects of certain securities (such as options), the convertibility of other securities to common stock, and executive bonus arrangements. The rules for disclosing earnings per share data can be encoded into an ES knowledge base. At year-end, the accountant invokes the ES inference engine, which asks the knowledge base and the accountant about potentially dilutive securities, contracts, and agreements and then recommends to the accountant the earnings per share that should be disclosed.

Auditing

The auditing standards generally accepted in the United States (ASGAUS), formerly known as generally accepted auditing standards, along with the statements on auditing standards (SAS) are issued by the Auditing Standards Board of the American Institute of Certified Public Accountants. They are guidelines—sometimes very broad, sometimes quite specific—on auditing procedures. Every audit must follow these standards. Auditors document their procedures in working papers. These standards (ASGAUS and SASs) contain a loose set of complex rules ideal for expert systems. Besides applying complex rules, another advantage of most ESs is they can make a risk assessment. In the interest of efficiency, most audits test accounts rather than proving all of them. Consequently, most audits and auditors are familiar with risk assessments; ESs have a remarkable ability to either make a specific risk recommendation or offer a risk range associated with the recommendations they make.

ASGAUS and the SASs identify four stages of an audit. Expert systems can be applied in each. Stage one is the acceptance or continuation of a client. Stage two is the planning phase. Stage three is the fieldwork. Stage four is completing the audit and issuing the report.

In the first stage of the audit, client acceptance or continuation, the standards include rules such as obtaining the client's permission to ask questions of the predecessor auditor. While the standards tend to be general, most certified public accounting firms and even internal audit departments now have their own specific rules for deciding whether to accept or continue with a client. These rules can be encoded into the ES knowledge base, along with the risks associated with the rule.

For example, SAS No. 82, Consideration of the Risk of Fraud, identifies a large number of fraud risk factors. Each factor alone presents only a limited risk of fraud, but when a number of risk factors are found together, the risk can be significant. These relative risks or ranges of risk can be associated within the ES knowledge base. An auditor moving through the fraud risk assessment or making inquiries of the predecessor auditor can invoke the expert system. The inference engine can make recommendations for follow-up activities and possible outcomes through the range of risks.

The three major activities that occur during the second phase of an audit, the planning phase, are (1) making preliminary judgment about materiality, (2) obtaining an understanding of internal control, and (3) establishing the audit risk. Expert systems can be useful in each. Most firms have fairly specific guidelines about establishing preliminary judgments about materiality and these guidelines can be coded into the ES knowledge base. The auditor decides which assertions are more at risk than others, the outcome of the fraud risk assessment, the users of the audit report, and any client-specific anomalies in the financial statements and provides this information to the ES. The ES communicates back the recommended materiality level for the financial statements as a whole and the amount of misstatement for each account and transaction class that is tolerable, along with risk levels or ranges.

ESs perform extremely well in assessing the control risk associated with clients. Clients answer internal control questionnaires and provide the results of the review of documentation of the internal control. These certain observations made by the auditor, and predetermined relationships about the adequacy of the design of the internal control system can be coded into the knowledge base. The recommendations solicited by the inference system about assessing control risk are communicated to the staff.

Some auditors use ESs to assess audit, inherent, control, and detection at the assertion level for each account balance and transaction class, coding the rules for these into the knowledge base. When the audit is being planned, the auditor invokes the inference engine and responds to queries that lead the expert system to recommendations about each risk for each account balance and transaction class. In addition to recommendations about different kinds of risk, the ES can also suggest specific audit procedures, the timing of the tests, and the extent of the tests. This can save valuable time.

During fieldwork, the third stage of the process, the auditor collects evidence to support an ultimate opinion, generally at the client location. One way ESs can contribute significantly to an audit is by providing guidance to new or lower-level staff members as they collect evidence. Rules for different approaches are coded into the knowledge base. A staff member who has a question or is uncertain about a given procedure invokes the inference engine for guidance on how to conduct the procedure and recommendations on how to interpret the results, as well as ideas for further procedures that might be appropriate.

Finally, the audit must be completed and an opinion report issued. In addition to final inquiries for potential contingent liabilities, completing the audit requires a search for subsequent events, a review of the working papers, and assessment of potential adjusting and reclassifying entries. Most firms have detailed steps for reviewing working papers that can be coded into the ES knowledge base, which can be invoked by the manager or partner at report-writing time. The ES responds with queries about the completeness of the tests or working papers and makes recommendations for dealing with missing and incomplete evidence, evidence that indicates significant problems, and other areas that may need further investigation.

The rules associated with the wording of auditor reports are intricate. An expert system that captures these rules can provide recommendations as to the appropriate type of report and suggested wording.

Income Taxes

Expert systems can be used in both tax compliance and tax planning, but while tax preparation firms realize significant results from using ESs, the volume and rate of change in tax rules precludes most organizations from gaining the benefit of a comprehensive ES for taxes because the cost of maintaining the knowledge base generally exceeds the benefit. Many tax provisions affect only a few people. But we'll give you an idea of how these systems work.

Tax Compliance

Generally, the designers of tax compliance ESs create a series oriented toward a particular tax form. Each rule associated with a particular form is coded into the knowledge base. When filling out the form, the inference engine asks the knowledge base about the tax rules and uses the responses given by the tax preparer to do calculations and make recommendations about how to report taxable income items.

For tax planning purposes, four additional factors are introduced into the process (1) alternative choices of treatment, (2) likelihood of being challenged by the Internal Revenue Service (IRS), (3) likelihood of an IRS audit, specific goal of the planning activity, and (4) likelihood of a change in the IRS rules during the planning horizon. Expert systems handle such uncertainties extremely well by factoring in the risks associated with each rule encoded in the knowledge base. When the tax planner enters a proposed transaction or event, the inference engine inquiry of the knowledge base produces the anticipated tax consequences and some indication of each of the risks associated with the choice. Using ESs to make these judgments saves significant time for tax planners.

Among the several expert systems used for tax planning are Deloitte and Touche's *World Tax Planner and ExperTax* from Price Waterhouse/Coopers. ExperTax uses a question and answer format to run a maze of 3,000 rules to outline a client's best tax options. Price Waterhouse/Coopers' also offers *COBRA* and *RIC Checklist* for tax compliance and *Tax Advisor* for estate planning, which is heavily tax-dependent.

NEURAL NETWORKS IN ACCOUNTING

There are an enormous number of opportunities to use neural networks in accounting, among them securities, accounts receivable, inventory, payroll, expense reimbursement, budgeting, performance measurement, and auditing. These are discussed below in terms of planning the NN application, identifying and selecting data, training the NN, using the NN, retraining the NN, follow-up procedures, and commercial software packages and services.

The superior pattern recognition power of NNs and the speed at which they operate offer significant opportunities to profit from their use in today's IT working environment. NNs generally are used in accounting to support decision making, to provide early warning of problems, and to confirm assessments.

Some NNs have been effectively integrated with other software applications to enhance data processing and to make decision recommendations. In these applications, NNs may function as a front-end, a middle, or a back-end process. Some institutions have found that NNs can rapidly isolate conditions in original data and signal an ES knowledge base (or other decision support module), reducing the time to access the search engine. This procedure is known as pattern recognition. In middle processing, the NN can take the results of a front-end process and signal the appropriate subsequent process. This process is called labeled filtering. In back-end processing, the NN receives input from other systems and makes a final decision. This process is called prediction. Accounting activities regularly use all three of these functions, which is why NNs offer opportunities to improve accounting activities.

Securities

The goal of managing securities is to use otherwise idle cash to contribute to earnings. This can be done in a vast range of ways, from maintaining cash in an interest-bearing bank account to purchasing and holding derivative securities. Often, investing idle cash is restricted to a limited class of investments to minimize risk while maximizing short-term return. NNs can be useful in two activities that contribute to this end (1) security selection and (2) daily monitoring of investment performance.

Selecting Securities

Depending on what the Board of Directors or other governing body has authorized, portfolio goals and objectives will narrow to some extent the selection of securities for investment. Once the CFO or controller identifies the general set of securities that meet portfolio requirements, NNs can be used to select specific securities to optimize the company portfolio.

An NN can assess the potential of a given security to yield the returns needed or to meet other investment objectives. Assessing both individual securities and different segments of the securities markets are the two main areas where accountants use NNs for portfolio management.

Planning the NN application. Designing an NN to assess individual investments requires identification of securities attributes that relate directly to portfolio objectives. In designing one to assess market segments, the goal is to predict trends and significant shifts to determine what action to take with respect to securities held in or being considered for the portfolio. Identifying a suitable measure for meeting the objective proves crucial to the successful use of an NN to manage a portfolio of securities. Both types of NNs are executed the same way.

Identifying and selecting data. The CFO or controller locates the data sources, generally a combination of internal databases and databases from subscription services. The plan should identify the required outcome of the NN and generally identify the data required—security, market segment, market data, the company's past experience with investments in those securities, and outcomes related to accomplishing objectives.

Training the NN for selecting securities. Results of past assessments are the best way to train a NN to assess individual securities. Otherwise, the NN is trained the same way other networks are trained (see pp. 63). In this case, the massive set would be all past experience with securities selected by the company and other allowable securities. The simplified set would be a representative sample of allowable securities and past outcomes.

Using the NN in selecting securities. The trained NN can be integrated with other decision support systems to monitor movements in the capital markets and initiate buy, sell, or hold decisions about securities held in the portfolio. By predicting upswings and downswings in the market, the NN can help maximize earnings on securities.

Retraining the NN. As the CFO or controller gains more experience with investments and market segments, the NN can be retrained by adding new experience to the training set. Most managers using NNs retrain the network every month or every quarter.

Follow-up and monitoring. The officer using a NN must constantly be aware of how well it is predicting how securities will perform and also how effective it is in helping achieve the goals of the portfolio investment strategy. If the network is only retrained once or twice a year and the performance of the selected securities does not conform to the NN predictions during any month or quarter, there should be a plan in place for retraining immediately.

Commercial software packages and services. BioComp Systems, Inc., offers *Profit 2000™*, which is described as an "end-of-day" financial market timing software. Profit 2000 helps time trading decisions to maximize profits by accurately predicting securities prices.

TradingSolutions from NeuroDimension, Inc., operates in a PC environment to simplify stock price forecasting.

Another PC package is *NeuroShell Trader* published by Ward Systems Group, Inc., which identifies buy and sell clusters of securities within a market.

MathWorks, Inc., offers *Datafeed Toolbox,* which connects with Bloomberg L.P. financial information services to download data to MATLAB®. The latter has NN, time series, and statistical functions for analyzing securities. This software can be integrated into any existing securities trading software.

State Street Bank has acquired a patent on a portfolio construction software package developed by State Street Global Advisors around a neural network. It uses NNs to predict the price movements of securities and helps accounting personnel integrate strategies with portfolio requirements.

Trading Securities

Most CFOs or controllers who manage a portfolio of securities formulate a strategy for the timing of buying, selling, and holding securities. Strategy in hand, they can then build NNs that monitor real-time data feed from the markets where securities maintained in the portfolio signal buy, sell, or hold recommendations based on trends in those markets.

As markets or specific securities reach buy/sell decision points, real-time data feed monitoring software notifies the portfolio manager a recommended course of action. The NN's rapid execution time enables the prediction model to operate in a fraction of a second, making adjustments to previous recommendations of just a few minutes ago. These predictions, based on the current price and most recent history, can be used to predict the price of a security thirty minutes from now, an hour from now, or on a future date. One researcher using a neural network was able to accurately (92 percent of the time) predict the price of gold futures thirty minutes from the current quote. Companies use real-time feed services to minimize losses and to capitalize on major price swings in a security price early in the fluctuation.

Planning the NN application. If the NN is to monitor specific securities or specific markets or market segments, the software may involve a series of neural networks, one for each security. Some up-front processing may be necessary to consolidate segment data prior to running the series. Lastly, software to communicate the results of NN predictions to the portfolio manager in real time basis must link to the neural networks. For that reason the NN is usually embedded in on-line or e-commerce software that accounts for securities.

Identifying and selecting data for trading securities. Data sources used vary but generally are a combination of internal databases, databases from subscription services, and direct feed from exchanges where portfolio securities are traded.

Training the NN for trading securities. An NN for trading securities is trained the same way other types of NNs are trained (see pp. 63). The massive set would be all past experience of securities held by the company and the simplified set would be a representative sample of securities and final outcomes. Don't forget to keep a holdout sample.

Using the NN in trading securities. The trained NN can be integrated with other decision support systems to monitor movements in specific securities and to initial buy, sell, and hold decisions. By monitoring changes in the market place, the neural network can predict the best timing for buy and sell decisions.

Retraining the NN. As the portfolio manager gains more experience with the securities markets, the NN can be retrained by adding new experience to the training set. NNs used for portfolio management may be retrained as often as daily because the securities markets are often volatile.

Follow-up and monitoring. The portfolio manager must be constantly evaluating market performance against the NN prediction. If the network is only retrained once or twice a month, and the selected securities fluctuate daily or weekly away from what was predicted, procedures should be in place for immediate retraining.

Commercial software packages. Lester Ingber Research provides consulting services to customize NN software for companies that trade commodities futures.

Accounts Receivable

CFOs, controllers, and accounts receivable managers use NNs to maintain information for granting credit to customers, to monitor sales, to detect fraudulent credit and collection transactions, and to make determinations about write-offs of customer accounts.

Credit Granting

Some entities extend credit to customers, clients, and others as part of their normal operations. These trade receivables do not require customers to provide collateral so it is important that companies extending credit do a thorough job of evaluating creditworthiness. Many companies, especially those with large volumes of accounts (large distributors, manufacturers with significant numbers of direct customers, and retail firms with their own credit cards), use NNs to flag credit applicants who may be poor credit risks.

Opportunities for using NNs include prescreening credit applications to eliminate obviously unworthy applicants and making final recommendations about granting credit. NN prescreening reduces processing costs and improves customer service because customers are given immediate feedback about credit decisions. Using NNs in the final approval process reduces analysis time and provides assurance about a professional's manual determination of creditworthiness.

The idea is both to reduce the time and cost of processing credit applications and to minimize losses by improving the quality of the credit-granting activity. Many entities have successfully provided trained NNs as part of the software available for those who make credit decisions even at the sales clerk level.

Planning the NN. Once again, planning begins with a clear and precise definition of what the network should accomplish, when it should do its work, and how the results should be communicated.

If the goal is to prescreen credit applications for extending trade receivables privileges, then the decisions about data, the type of NN, and the outcome can be stated succinctly. Once information from the application form is entered, the computer accesses credit bureau reports and other stored data and then makes the creditworthiness determination. This way, though, the company must pay for the processing of the application, including a fee to the credit bureau. If there is sufficient data in the original application, the company can avoid many of these processing costs by using a NN to make the final credit decision.

Manufacturers and wholesale distributors use NNs to help them keep up with changing creditworthiness profiles for their customers. With large data sources for large customers, NNs are excellent at accurately identifying relationships among the data that can affect the credit profile.

Identifying and selecting data. Applications for credit often contain several pages of data as well as supplementary financial statements. In general, the more data, the better for the NN. Generally speaking, an NN need data from both successful and failed credit experiences. If the entity has no data available on unsuccessful trade credit accounts, it will be impossible to train a NN. Using a small sample (100 to 150) of accounts in several trials or pilot projects and varying the number of inputs can help identify the best combinations of inputs for NN development. Also, use as much data as is available over the most recent history, from the last two quarters to the last five years of data. The date of application may also provide important data about economic conditions external to the account that might affect creditworthiness.

Codes that use alphabetical characters must be converted to either a cardinal or ordinal scale. In applying a cardinal scale, leave enough space between numbers to allow for discrimination between observations (i.e., if four geographical regions have different economic trends, instead of numbering them one, two, three, and four, use two, four, six, and eight).

Training the NN application. This type of NN is trained the same way other types of NNs are trained (see pp. 63). The massive set is all past experience with trade accounts receivable customers of the company and the simplified set is a representative sample of good and bad trade accounts receivable.

Using the NN in granting credit. The trained NN can be installed as a subroutine within any of the operational software used by the entity to make credit decisions. For prescreening, it can be executed at the end of the data entry program or at the end of the input just after all the validity check routines have been executed. That way a customer applying at a division or on-line can be notified immediately of the credit decision.

If the goal is a general assessment of creditworthiness to establish or adjust a major credit limit, the NN is embedded in the software near the end of processing the application, after all inputs have been obtained. There, the network usually makes a recommendation to the CFO, credit manager, or sales manager, who makes the final determination, though sometimes the network makes the credit granting decision.

Retraining the NN. As the company gains more experience with trade credit losses and as credit profiles change, the NN can be retrained by adding new experience to the training set. This is best done at least monthly or quarterly.

Follow-up and monitoring. Companies using NNs should have a procedure for analyzing why accounts are written off so that they will be aware if the network is failing to detect non-creditworthy applicants. Whenever the rate of account write-offs increases significantly during any month or quarter, follow-up procedures should call for immediate retraining.

Commercial software packages and services. HNC Software Inc. offers *Capstone™ Decision Manager,* which can easily be integrated into the application processing system of any trade receivable decision environment. *Capstone Online* is an HNC-hosted ASP solution that makes automatic credit decision-making available to companies. The software connects with point-of-sale machines as well as major consumer credit and business reporting agencies.

Auditor, developed by C. Duncan (University of Illinois), helps in the analysis of the allowance for bad debts.

Authorizer's Assistant is used by American Express and others for credit evaluation and authorization.

General Motors' *Credit Advisor* appraises customer credit using a scoring system.

The Edge™ is a system from Neuristics, LLC that can be integrated into existing company credit granting software that promises to identify good credit risks that have been traditionally misclassified by standard credit scoring tools. It's primarily designed for identifying prospects in the subprime market. The company also sells *Nexcore™* as the next generation of NN-based credit risk scoring. Nexcore is designed for all types of credit scoring.

Detection of Credit Fraud

One of the more common forms of fraud is theft by obtaining goods and services with trade credit with no intention of repayment. These thieves take advantage of bad debt experiences of companies that normally arise from competition and the vicissitudes of the marketplace. As more merchants use more on-line authorizations of credit purchases, thieves regularly apply for immediate credit. An NN can monitor the system to identify potential fraudulent credit applications, giving the company an opportunity to refuse authorization of immediate purchases.

Planning the NN application. This type of NN is designed to recognize patterns that indicate the possibility of fraud. Because they know at some point the company will discover the fraud, for instance, thieves attempt to charge up the entire credit limit very quickly, usually for goods that can be readily resold. This results in subtle patterns of purchase that NNs are very good at detecting as credit sales authorizations are called in or received on-line. This type of pattern-recognition software can be embedded in a company's accounts receivable and sales authorization software of the company.

Identifying and selecting data. Data for setting up this type of NN may include past purchasing and payment trends of trade accounts receivable, frequency of transactions, size of each transaction, characteristics of customers (from the trade credit application), and other account data. Data from accounts that have been fraudulently submitted should be kept with similar data from good accounts. Companies need as much data as possible, usually two to five years of history.

Training the NN to detect fraud. This type of NN is trained the same way other types of NNs are trained (see pp. 63). The massive set is all past experience of trade accounts receivable transactions and the simplified set is a representative sample of both good and fraudulent accounts receivable transactions. Don't forget to keep a holdout sample for testing the new NN. This test provides assurance that the network can accurately recognize fraudulent use of trade credit.

Using the NN for credit fraud detection. The trained network can be installed as a subroutine within any operational software used to process credit sales authorizations, near the end of processing the sales authorization, after all inputs have been obtained.

When the NN is used to help prevent fraudulent use of trade credit cards and to recover stolen cards, it is embedded within the software used to authorize credit card purchases. When a pattern emerges to indicate potential fraud, the software notifies the sales clerk seeking authorization to confiscate the card and contact the company. Meanwhile, the company can contact the legitimate owner of the card.

Retraining the NN. As the company gains more experience with credit fraud losses, the NN can be retrained by adding new experience to the training set. This should be done at least monthly or quarterly so that the system reflects the most recent changes in experience as well as the historical experience.

Follow-up and monitoring. Companies using NNs should have procedures for monitoring how well the networks are performing in detecting fraudulent use of trade credit and for follow-up on any anomalies with immediate retraining.

Commercial software packages and services. The BioComp Systems, Inc. *Financial Systems* package includes an NN-based filter to detect customer fraud.

ACI Worldwide offers several products marketed under the brands *OCM24®, TRANS24®,* and *i24™.* These products, which facilitate e-payment, include an NN-based e-fraud detection and management service for debit and credit card issuers as well as merchants.

PRISM CardAlert from Nester, Inc., is a fraud risk management system that has been praised for reducing fraud losses over the years.

Oracle Corporation's *Oracle Special Investigative Unit Support System* aids in fraud investigations by detecting hidden patterns within large volumes of data.

HNC Software Inc. offers *Falcon™* and *Falcon™ Model 2000,* credit card transaction fraud detection packages that are used by a number of financial institutions. The company's *HNC Credit Intelligence Solution* is designed to minimize credit risk, reduce fraud, and improve customer service. Its *Eagle™* is a risk management system designed to protect vendors from emerging fraud trends.

TransAlliance LLP offers *Cardholder Risk Identification Service* to detect debit card fraud. Many credit unions use the service. In addition to early detection and prevention of fraud losses, users claim reduced insurance premiums for insurance.

Anthem Corporation's *Financial Crime Investigator* helps identify fraud in purchases and contracts.

Monitoring Credit and Writing Off Bad Debts

Companies periodically write-off the unpaid balances of uncollectible trade receivable accounts. When trade credit customers become financially stressed, the patterns of data (periodic financial disclosures, purchases, and payments) associated with their accounts begin to change. Determining that a customer is financially stressed before the customer goes into bankruptcy may reduce losses significantly. Neural networks are adept at recognizing the pattern of purchases and payments associated with financial stress, along with relationships in financial disclosures, early in the process so that the company can take action to prevent losses.

Some companies use NNs to monitor purchase and payment activity on all receivables to provide an early warning when action may be advisable before the account fails to pay or bankruptcy otherwise threatens. Others use the networks to signal when a debt is no longer collectible. Some use NNs for both.

Planning the NN application. Though planning and design is slightly different depending on whether the goal is monitoring or predicting bankruptcy, the procedures for executing the NN are similar. For monitoring, the design requires identification of attributes that relate directly to recognizing when customers are becoming financially stressed, such as changes in the size and frequency of purchases and payments over time. For recognizing financial failure, attributes of interest might be the periodic financial disclosures compared with purchase and payment history.

Identifying and selecting data. The CFO, controller, or credit manager identifies the attributes needed to develop the NN for the required outcome. Data sources are generally a combination of company internal databases and databases from subscription services and credit bureaus, together providing information on financial situation and purchases and payment history with the company and elsewhere.

Training the NN for monitoring accounts receivable. This type of NN is trained the same way other types of NNs are trained (see pp. 63). In this case, the massive set is all financial disclosures and transaction history of trade accounts receivable and the simplified set is a representative sample of financial disclosures and transaction history from both good and bad trade accounts receivable.

Using the NN to predict and close out bad debts. The trained network can be integrated with other decision support systems to monitor, daily or weekly, the behavior of trade credit receivable accounts of the company and initiate extension or retraction of credit, schedule a workout, or write off an account.

Retraining the NN. Over time the patterns of customers' financial disclosures, purchases, and payments also change. As the company gains more experience with good and bad trade credit accounts, the NN can be retrained by adding new experience to the training set. Doing this at least monthly or quarterly is recommended.

Follow-up and monitoring. As with all types of neural networks, companies that use them for predicting trade credit account behavior need procedures to check that the NN is performing. Although there will always be some bad debt, there should be no bad debt surprises. If there are, procedures for retraining the NN should be followed immediately.

Commercial software packages. ContiAsset Receivables Management, a service vendor, uses neural network software to predict whether accounts will go bad.

Applied Analytic Systems, Inc., provides both software and consultation to implement NN software.

The BioComp Systems, Inc. offers *Financial Systems* software that includes NN-based methods for identifying financially stressed customers.

Inventory Accounting

CFOs, controllers, and inventory managers use neural networks to manage and value inventory with electronic data interchange (EDI) and electronic funds transfer (EFT) along with point of sale (POS) systems, fewer paper documents are available for these purposes. Immediate update of records is an advantage of on-line inventory systems, but independent physical counts are too infrequent to provide effective control. NNs provide control over inventory by recognizing patterns that suggest fraudulent reporting, inefficiencies in processing, and cover-up of inventory theft. They are also efficient at predicting changes in inventory prices and movement of inventory to help prevent inventory obsolescence and identify items that may need cost or market adjustments.

Inventory Control

Internal controls are necessary to assure the accuracy of inventory reporting. When inventory recordkeeping is computerized, the data is readily available for incorporation into neural networks. The NNs then monitor inventory transactions early to identify potential errors and fraud in the automated inventory system so that management can take action to minimize their effects. If inventory is counted and priced only quarterly or annually, errors or fraud may not be discovered until months after the fact, meaning that interim reports may be materially misstated or frauds continued over a long period of time.

Planning the NN application. This type of neural network is designed to recognize data patterns associated with erroneous and fraudulent inventory transactions. Errors in computerized systems generally arise usually because staff members have misapplied procedures for capturing input and more rarely from programming errors. Employees who steal generally provide input data that seems convincing, but transaction data from false inventory transactions often has a pattern that, though subtle, can be recognized by a neural network.

This NN application is designed to be embedded in inventory posting software. The NNs can be customized to alert inventory managers to potential problems and the need for immediate investigation and perhaps selected physical counting.

Identifying and selecting data. The data to be selected for inventory control include past inventory transactions for a given inventory, frequency and size of transactions, characteristics of customers (for inventory sales transactions) and of suppliers (for purchase transactions), and employee activity. It should include data from past inventory accounts that have been subjected to fraud or errors along with similar data for good transactions. The more data the better—at least two years of history and preferably five.

Training the NN. This type of NN is trained the same way other types of NNs are trained (see pp. 63). The massive set is all experience with all inventory transactions and the simplified set is a representative sample of good, erroneous, and fraudulent inventory transactions. Once the network is trained, use a holdout sample to test it.

Using the NN to detect inventory errors or fraud. The trained network can be installed as a subroutine within any operational software used to process inventory transactions near the end after all inputs have been obtained. Usually, the NN generates a message to the inventory manager and controller when additional investigation seems advisable.

Retraining the NN. Thieves are ingenious, so over time, while patterns of errors may change, patterns of fraud in inventory transactions certainly will. As the company gains experience, the NN network should be regularly retrained, at least monthly or quarterly, by adding new experience and the benefits of new insights to the training set.

Follow-up and monitoring. NN results should be compared with actual experience to make sure the neural networks are successfully detecting erroneous and fraudulent inventory transactions. In case errors and fraud are regularly discovered without the help of the NN, procedures must be in place to retrain the network immediately.

Commercial software packages and services. Applied Analytic Systems, Inc., provides both software packages and consultation on implementing neural networks.

DynaMind also sells NN-based fraud detection software solutions.

Inventory Valuation

For pricing purposes, inventory items are classified as (a) "cost less than market price" (which do not need adjustment), (b) "market price less than cost" (which do), and (c) obsolete inventory (which need to be written off). When preparing balance sheets, controllers must identify obsolete inventory as well as inventory items whose prices need to be adjusted because the market price is lower than the cost. This procedure can take significant effort and may delay the closing process significantly. However, NNs can simplify the task by quickly identifying those inventory items that need adjustment. This reduces inventory time and cost, because staff can then concentrate on proper valuation. Some companies find NNs particularly useful when they have a number of branch locations. They make a trained NN part of the software available for all their inventory clerks and cost accountants.

Planning the NN. Once the company has a clear and precise definition of what this type of network should accomplish, the goal is to identify inventory items with a high potential for diminished economic value or obsolescence. When items are no longer selling, the purchase price has been declining, bundles of related items are discontinued, or other items can be substituted, the potential for obsolescence or for cost exceeding the market increases dramatically.

Identifying and selecting data. Transaction data for inventory items vary significantly depending on the type of purchase and sales processing software the company uses. As usual, the more data available to the NN, the better. In general, categories of inventory items like those with a cost less than market, a cost greater than market, non-obsolete, and obsolete should be used. Without data on obsolete inventory or inventory with costs that should indicate the need for a write down, it will be impossible to train the network. Pilot projects using a small sample of inventory items from each category (100 to 150) that vary the number of inputs to the network can help identify the best combinations of inputs for NN development.

In general, to be effective, the NN needs the history from the last two quarters all the way through the last five years, including the dates of purchase and sale. The latter can provide important data about external economic conditions that might affect the classification of an item. With this much data, often codes use alphabetical characters. These must be converted to either a cardinal or ordinal scale, making sure the scale chosen makes it possible to discriminate easily between observations.

Training the NN for predicting inventory write-downs. This type of NN is trained the same way other types of NNs are trained (see pp. 63). The massive set is all inventory items; the simplified set is a representative sample of inventory items including cost less than market price, market price less than cost, non-obsolete, and obsolete. The system should be tested with a holdout sample.

Using the NN. The trained network can be installed as a subroutine within operational software near the end of processing the inventory subsidiary ledger, after all inputs have been obtained. Usually, the NN makes predictions to the inventory manager every week or month and to the controller just before closing or when inventory pricing adjusting entries are being prepared.

Retraining the NN. As the company gains more experience with managing the pricing of inventory categories electronically, the neural network can be retrained by adding this new experience to the training set. Retraining monthly or quarterly is advisable.

Follow-up and monitoring. It's wise to keep an eye on market price fluctuations externally and analyze the reason for the changes and for obsolescence in inventory items to test whether the network is accurately classifying items by price category. Any time the network is noticeably not performing as expected during any month or quarter, the observation should be followed up with retraining according to a planned procedures.

Employee Expense Reimbursement

Employees are entitled to reimbursement for expenses they have incurred on company business, but it is not unheard of for employees to file fictitious or exaggerated expense reimbursement claims or claims for expenses incurred for personal purposes.

False reimbursement claims are in fact an all-too-common form of employee fraud. A neural network can monitor expense reimbursement claims to identify potential fraud and allow the company to refuse to authorize payment or flag claims needing further manager attention.

Planning the NN application. This will be a pattern-recognition type of NN. Although employee thieves attempt to make their claims convincing, data from false claims often has a pattern that though subtle can be recognized by a neural network.

The NN is designed to be embedded in the company's reimbursement processing or voucher processing software and alert managers to the possibility of employee expense fraud.

Identifying and selecting data. The data used in this type of NN may include past employee expense reimbursement claims for a given activity, frequency of claim transactions, size of each claim, and characteristics of employees as well as other data. Data from fraudulently submitted expense accounts are as important as valid claims data. Companies usually use two to five years of history.

Training the NN. This type of NN is trained the same way other types of NNs are trained (see pp. 63). The massive set is all employee expense reimbursement transactions and employee databases and the simplified set would be a representative sample of good and fraudulent employee expense reimbursement transactions and related employee databases.

Using the NN to detect fraudulent employee expense claims. The trained network can be embedded as a subroutine in the company's software near the end of process for authorizing expenses and preparing vouchers, after all inputs have been obtained. It usually generates a message for the manager and controller when additional investigation or other follow-up is needed.

Retraining the NN. Patterns of fraudulent employee expense reimbursement claims change over time. As the company gains more experience with fraudulent expense claims, the NN can be retrained by adding new experience to the training set. Monthly or quarterly retraining is recommended.

Follow-up and monitoring. As usual, NN performance must be compared with any fraud companies detect without the neural network. In case it's discovered that the network is not detecting frauds well, the company must have follow-up procedures in place for prompt retraining with new information.

Commercial software packages and services. Strategist from Unitek Technologies' analyzes expenses for reasonableness. It looks at financial statements and reveals the effect of different actions on the financial ratios.

Payroll Control

In some companies payroll (certain contractors and professional service organizations) is directly tied to revenues generated by the firm. Controllers must be confident of the accuracy of payroll data and payroll processing. Most companies maintain computerized time-keeping and payroll processes. Errors and fraud in the capture of time and activity, as well as an occasional programming error, can affect the accuracy of payroll accounting. Neural networks can monitor payroll transactions early to detect any erroneous and fraudulent transactions so management can take action to minimize the negative effects of the errors or frauds. As with billing customers and pricing inventory, if payroll transactions are only reviewed quarterly or annually, there can be serious consequences for the financial reporting function.

Planning the NN application. This, again, is a pattern recognition process. Errors in computerized systems generally arise because of staff mistakes in the capture of input, though occasionally there are programming errors. Employee thieves who benefit from these mistakes or who intentionally create fictitious transactions are generally aware that at some point the company may discover the fraud and try to build their fraud around data that seems convincing, but an NN can detect the subtle pattern of a false or erroneous transaction.

This type of NN can be embedded in payroll preparation or distribution software to alert managers that something needs to be checked out.

Identifying and selecting data. Data used in this type of NN may include past payroll transactions for employees, frequency and dollar amount of transactions, job classifications, human resources data, and certain production and activity measurements. The company needs data from fraudulent or erroneous employee records as well as data for good transactions. Usually two to five years of history is appropriate.

Training the NN. This type of NN is trained the same way other types of NNs are trained (see pp. 63). The massive set is all payroll transactions and related employee databases; the simplified set is a representative sample of good, erroneous and fraudulent payroll transactions, and related employee databases. Do not forget to save a holdout sample to test the ability of the trained NN to do its assigned job.

Using the NN to detect payroll error or fraud. The trained network can be installed as a subroutine within operational software near the end of processing the payroll, after all inputs have been obtained. The NN generates a message for someone like the human resources manager, the timekeeper, or the controller to investigate immediately.

Retraining the NN. Staff can cause errors in a variety of ways. They can also come up with an ever-changing variety of ways to commit fraud. As the company gains more experience with erroneous and fraudulent payroll transactions or learns more about what is happening elsewhere, the neural network can be retrained by adding this new experience to the training set. Retraining should be done at least quarterly, preferably monthly.

Follow-up and monitoring. If there is a significant increase in the rate of erroneous or fraudulent payroll transactions detected by means other than the NN, the company should have in place a procedure for prompt retraining of the NN.

Budgeting

CFOs and their colleagues need a variety of prediction tools to create accurate budgets while dealing with the inevitable budgetary hedges operational and support unit managers insert during the annual budgetary exercise. If there is too much hedging, the budget is unrealistic and executive decisions about strategies for the coming year cannot be achieved. The possible result is significant losses for the company. Identifying excess hedging may not be easy because of the many details incorporated into budgets. Neural networks can quickly identify excessive hedge or budgets that are unrealistic for volume or other considerations.

The budget director must review each account from the lowest level up to the consolidated entity, budget by budget with supporting detail, for appropriateness. A neural network can prescreen accounts by unit and identify those that are most likely to be over-hedged. This will reduce the time and cost of the budgeting process as well as identifying unrealistic projections that could undermine company planning.

Planning the NN. Once the goals of the network are precisely defined, the decisions about data, the type of neural network, and the outcome can be stated succinctly.

Identifying and selecting data. Transaction data needed for budgeting vary significantly by budget system and performance reporting system, but generally speaking, the company needs historical budget and actual data from each budgetary unit along with additional performance measures. The network learns by showing the network historical data for each account both budgeted and actual, with an indication of which accounts have been excessively hedged and which accounts are reasonable. Without data on budget and actual results for each unit, it will be impossible to train a neural network. This is another area where using small pilot projects of 100-150 samples and varying the number of inputs will generate the best combination of inputs to incorporate into the NN. The network will in any case need history from the last two quarters through the last five years of data.

Training the NN to support budgeting. This type of NN is trained the same way other types of NNs are trained (see pp. 63). The massive set is all budget and actual experience for all accounts and a simplified set is a representative sample of budget and actual experience for both reasonable and excessively hedged units.

Using the NN to detect excessive budgetary hedging. The trained network can be installed as a subroutine within the company's budgeting software near the end of processing the budget, after all inputs have been obtained. The NN alerts the budget director to those accounts that seem to be hedging excessively.

Retraining the NN. As both internal company and external economic circumstances change, so does budgeting. As these changes occur and as the company gains more experience with detecting excessive hedging, the NN can be retrained by adding this new experience. Monthly or quarterly retraining is recommended.

Follow-up and monitoring. If budget-to-actual experience makes clear that there has been excessive hedging and the NN has not caught it, there should be procedures in place for immediate retraining of the network.

Commercial software packages and services. Neural Ware Inc. offers *Neural Works Predict.*

Performance Measurement

CFOs and controllers must measure financial performance against other operational performance to evaluate how the company is doing. Systems used to capture data for financial measures are different from those used to capture data for operations measures. Moreover, timing differences in the source data and procedural differences in data adjustments (accruals, reversing entries, and other adjustments that require estimates) may cause significant inconsistencies between systems measuring financial performance and operations performance. NNs can help identify these inconsistencies before they turn up in reports, embarrassing for the financial and operations executives responsible for the conflicting departments.

If inconsistencies are not identified before they are communicated to management and the board, the credibility of the financial reporting system is jeopardized. However, once NNs promptly identify inconsistencies between operational and financial performance measures for a given organizational unit, staff can review the inconsistent information and be prepared to explain it.

The large number of managers and information systems used to capture performance data from different organizational units can geometrically magnify problems of inconsistency between budgeted and actual amounts. Generally, the preparation of the combined reports occurs with limited time for review and makes detailed examination of all units impossible.

This is a situation in which NNs perform particularly well. They reduce the time and cost of reviewing details supporting operational and financial performance reports.

Planning the NN. The decisions about data, the type of neural network, and the outcome must be stated succinctly, with special attention to direct relationships where certain financial accounts should change, directly or inversely, if certain activities increase or decrease.

Identifying and selecting data. Much depends on the type of software used to prepare performance reports. In general, the more data it has available from performance report databases, the better the NN can operate.

Generally speaking, the company needs to have operational and financial data that is known to be consistent, along with similar data that is known to be inconsistent. In small pilot tests on perhaps 100 to 150 accounts, by varying the number of inputs to the network, the company can identify the best combination of inputs for the NN under development. In general, there should be as much data as possible over the history from the last two quarters to the last five years. Again, the dates of reports may be important to give the NN insight into cyclical economic or other market conditions that might cause a unit's performance in a given period to be classified as inconsistent.

Wherever there is a large volume of data, as in this application, preparing the data may be burdensome. Because neural networks require numerical data representation, codes that use alphabetical characters must be converted into numerical scale that allow for the appropriate level of discrimination between observations.

Training the NN to detect performance inconsistencies. This type of NN is trained the same way other types of NNs are trained (see pp. 63). The massive set is all financial and operational performance reporting data. The simplified set is a representative sample of both consistent and inconsistent financial and operational performance reporting data. As usual, once the network is trained, it should be tested with a holdout sample.

Using the NN. The trained network can be embedded as a subroutine within the operational software used by the company to prepare the final performance report, to be run near the end of the process, after all inputs have been obtained. Usually when the NN recognizes inconsistency in the financial and performance data for a given unit, it will send a message to the CFO to investigate further.

Retraining the NN. Changing company and economic circumstances can change the patterns and the relationships between operational and financial reports. To be sure the NN is responding to such changes, it should be retrained monthly or quarterly by adding new experience to the training set.

Follow-up and monitoring. If managers identify inconsistencies between operational and financial performance measures that have not be detected by the NN, the network must be retrained immediately using procedures already planned for.

Commercial software packages and services. Triant Technologies offers *Model Ware.*

Auditing

Auditing is assessing evidence to form an opinion about the accuracy of financial statements. Neural networks can be of significant help in confirming relationships in financial statements and identifying and providing other evidence to auditors. Specifically, NNs can support internal control assessments, assessments of potential fraud and operational audits, and can generate evidence for the going concern assumption.

Internal Control Assessments

AICPA Statement on Audit Standards No. 78 requires that auditors understand enough about client internal controls to plan and conduct the audit. NNs can recognize patterns of internal control and make a recommendation on control risk to be used in audit planning. They can quickly identify the control risk associated with various internal control components. Usually, an auditor will use professional judgment in making an assessment and then use the NN to confirm it. This enhances the quality of the audit.

To learn more about internal controls auditors use questionnaires, observation, and client documentation among other evidence. Many auditors input this evidence into audit software in their automated working papers, which the NN then processes to make a recommendation about how much control risk to assign in the audit. The goal of using NNs is to reduce the time and cost of making internal control assessments and to improve their quality.

Planning the NN. Planning is fairly straightforward once the auditor defines precisely what the neural network is to do—determine and communicate the control risk of a given client.

Identifying and selecting data. Though the data used to assess control risk varies from auditor to auditor, it generally includes results from internal control questionnaires and other client inquiries, direct observations by the auditor, and client documentation, all entered into the audit software, and the more the better. The network learns from historical data about items entered as part of the computerized working papers that are labeled with control risk assessments (i.e., maximum, very high, high, moderate, low, very low). Without computerized data on previous control risk assessments, it will be impossible to train a neural network. Pilot projects of small samples that vary the number of inputs are useful in finding the best combination to use in the NN. The data should cover the client's history from the last two quarters through the last five years.

The date of each assessment can give the NN important data about certain external economic conditions that might affect control risk over time. Also, NNs require numerical data representation, which means that codes using alphabetical characters must be converted.

Training the NN. This type of NN is trained the same way other types of NNs are trained (see pp. 63). The massive set is all internal control risk assessment experiences for all clients. The simplified set would be a representative sample of successful and unsuccessful control risk assessment experiences. There should be a holdout sample for testing the NN after it's trained.

Using the NN to predict control risk. The trained network can be installed as a subroutine within the audit software embedded near the end of processing initial inputs related to understanding the internal controls for a given client. Usually, the NN relays its prediction about control risk to the auditor in the field so that initial planning can begin promptly.

Retraining the NN. Over time the profiles of internal control assessments change, and as the auditor gains more experience with managing control risk for more clients, the NN can be retrained by incorporating these new experience into the training set. Most auditors retrain their NN monthly or at least quarterly basis.

Follow-up and monitoring. It doesn't take long to find out whether NNs are predicting control risk well. If the NN's control risk assessments differ from what control testing evidence reveals, the auditor must analyze the differences and promptly institute predetermined procedures to retrain the NN to properly predict control risk. More frequent scheduled retraining usually makes this unnecessary, however.

Commercial software packages and services. **TICOM,** developed by A. Baily and M. Gayle, evaluates internal control systems.

Internal Operations Risk Analysis Software from Business Foundations appraises a company's areas of risk and its internal control structure on the basis of answers to about 200 interview questions, and examines operational strengths and weaknesses. Risk is assigned to categories. It evaluates working environment, objectives, planning, and personnel.

Fraud Detection

AICPA Statement on Auditing Standards No. 82 requires auditors to assess the potential for material frauds by management misstating financials or employees misappropriating assets. Unfortunately, because of the design of their compensation packages, managers are often tempted to overstate sales or understate expenses by generating fictitious transactions on their financial reports. These require particular scrutiny from the auditor. Similarly, client personnel have been known to steal from their employers and create fictitious transactions to cover up the theft.

The patterns created by both types of fictitious transactions, though subtle, can be recognized by a NN even though it would be missed by manual review of financial statements or many other types of analytical procedures.

Planning the NN application. Again, we have a pattern-recognition type of neural network, one that looks for relationships of transactions and account balances that signal the possibility of fraud. Employees (whether management or staff) will attempt to provide sufficiently convincing backup for the fraud cover-up. This results in subtle patterns of transactions and account balances that NNs can be trained to detect.

Identifying and selecting data. What this type of NN needs is transaction data and account balance data for clients, frequency and dollar size, dollar size of transactions, and other data characteristic of the client. The NN needs two to five years of history of fraud as well as clean books.

Training the NN. This type of NN is trained the same way other types of NNs are trained (see pp. 63). The massive set is all transaction data from all clients; the simplified set is a representative sample of good and fraudulent transaction data from the massive set. There should also be a holdout sample on which to test the network once it's been trained.

Using the NN for fraud detection. The trained network to flag potentially fraudulent transactions can be embedded as a subroutine within the audit software near the end of processing for client transactions but before sampling transactions for tracing and vouching. When a pattern emerges that suggests fraud, the software notifies the auditor to adjust the audit program to investigate the transactions flagged.

Retraining the NN. Patterns of fraud in transactions and account balances change constantly. As the auditor gains more experience with and understanding of fraud, the neural network can be retrained by adding this new experience to the training set. It is recommended that networks be retrained at least quarterly but preferably monthly.

Follow-up and monitoring. Auditors must be constantly watching to make sure the NNs are actually detecting fraudulent financial transactions. If there is significant variance from what might be expected, procedures should be in place for immediately retraining the network.

Commercial software packages. PRISM CardAlert from Nester, Inc., is a fraud risk management system that has been praised for reducing fraud losses over the years.

KPMG's *Inherent Risk Analysis* and Price Waterhouse/Cooper's *Risk Advisor* are similarly appropriate for use in risk analysis.

Assessing the Going Concern Assumption

At the end of audit fieldwork, the auditor must evaluate the going concern assumption for the client. When a client becomes financially stressed, the pattern of purchases and payments begins to change, as do the relationships of account balances on the balance sheet. Determining that a client is financially stressed with the possibility of going into bankruptcy helps the auditor decide if it is appropriate to add a "going concern" warning in the audit opinion.

Neural networks are adept at recognizing worrisome patterns of purchases and payments and troubling account balances early in the process. This information not only gives the auditor warning of trouble, it provides evidence to support the warning. The auditor can then make recommendations to management about necessary actions before it is too late. The evidence will also support the auditor's decision to include a going concern paragraph in the audit report. Thus, neural networks can be used to identify clients that are likely to go into bankruptcy, or fail financially.

Planning the NN application. Planning and design of the neural network will vary slightly depending on whether it is to be designed to analyze account balances only or both transactions and account balances, but the procedures for executing both variations are similar. In the second, dual-use, type of NN, the design will identify such attributes of the pattern as change in the size and frequency of sales, purchases, and payments over time for individual accounts as well as the periodic financial and non-financial disclosures by the client that are needed by the first type of NN.

Identifying and selecting data. In designing an NN to assess going concern status, the auditor generally selects a combination of the client's databases and databases from subscription services and credit bureaus. Data selected include sales, purchases and payment transaction history, financial disclosures, industry data, and credit bureau information.

Training the NN for going concern evaluation. This type of NN is trained the same way other types of NNs are trained (see pp. 63). The massive set is all data from all clients; a simplified set is a representative sample of good client data and data from bankrupt companies or clients that entered into receivership. A holdout sample makes it possible to test the NN's accuracy after it has been trained.

Using the NN in the going concern evaluation. The trained network can be integrated with other audit software to assess clients of the audit firm and communicate to the auditor a warning about any necessity to give a going concern warning about the client in the unqualified audit report. By evaluating daily and weekly client activity, the NN can predict financial stress early enough so that the auditor can alert client management to the threat of bankruptcy.

Retraining the NN. As the auditor gains more experience with evaluating the going concern status for more clients, the neural network can be retrained by adding new experience to the training set. Retraining at least monthly is preferable to retraining at longer intervals.

Follow-up and monitoring. Just as accountants monitor actual versus budgeted expenditures, auditors must monitor actual versus predicted going concern problems to determine how well a NN is performing in predicting financial stress in clients. If there is a divergence and client performance differs noticeably from what the NN predicted, it should be retrained immediately following preset procedures.

Commercial software packages. HNC Software, Inc. offers the *ProfitMax Bankruptcy* software package to detect the likelihood that a company will go into bankruptcy.

Operational Auditing

Operational audits, usually run by internal auditors, are often conducted regularly as a management tool. Their goal is to find ways to improve operational effectiveness. With the increasing use of information technology, significant databases have been developed to track operations. Auditors can apply NNs to these databases to mine data, enhance processes, and monitor control procedures. In mining data, for instance, they can identify processes on which it might be productive to perform an operational audit.

Data-Mining. The amount of data available about operations is often so vast that it conceals significant information companies could use to improve their efficiency; this is the situation called "information overload." NNs can be used to rapidly identify large as well as small clusters of operations with high or low potential for productivity improvements.

Process Enhancement. Operational auditors must make recommendations for process improvements as an audit outcome. NNs can suggest opportunities for process enhancement across a broad range of processes from agriculture and mining through manufacturing and distribution to service and e-commerce processes. Consequently, all company processes may be candidates for use of NNs to improve their effectiveness.

One goal of using NNs for process enhancements is to reduce the time and cost of adapting to changing conditions in throughputs. These new conditions may result from changes in inputs to an operation, in process resources, in the operations environment, or in output. NNs can instantaneously recognize changes in an operation. They read sensors that monitor operations and signal changes that can be made to optimize throughput.

Control Procedures. Operational auditors usually follow up to see if management implemented their audit recommendations. They also examine whether the recommendations resulted in the expected improvements. Sometimes, follow-up monitoring can be accomplished through neural networks. NNs simultaneously measure the quality and quantity of a process's output and indicate deviations from expectations. NNs can also be used as a form of quality control inspection, which can under the best of circumstances significantly enhance process yields. The NNs monitor throughput and signal when processes should be fine-tuned to continue to produce a high yield.

The goal of using neural networks as control procedures is to alert management when multiple changes in operations activities, each within a tolerable range, collectively present a threat to the effectiveness of the operation. It can often happen that throughput of an operation is suboptimal due to interaction effects of the many activities that comprise it. Even if one activity is at the high end of its tolerance range, it can put undue stress on a related activity that is operating at the low end of its tolerance range. NNs can deduce immediately when combinations of tolerable operational activities will result in a lower process yield than is possible and communicate to management to take action quickly.

CHAPTER 6

Legal Applications

Lawyers use computers extensively. Until recently, however, their primary use of computers was either to automate office tasks or to perform research using electronic databases. Developments in AI technology will ultimately allow computers to emulate substantive legal work—allowing computers to reason like lawyers. While AI technology applications to the legal field have not yet reached the level of fully automated expert systems, the trend in that direction is clear. The AI techniques that can be used are case-based reasoning, fuzzy logic, and neural networks.

So far, AI in the legal field has been concerned primarily with two types of systems:

1. General legal expert systems for performing research and other tasks.

2. Modeling legal reasoning for predicting the outcome of litigation.

In deciding what action to take, lawyers apply specific facts through a series of reasoning steps. The large number of subareas of law makes this an overwhelming task, even for lawyers who specialize in a given subfield. And lawyers have to deal with numerous statutes, which are often reinterpreted in decisions by the appellate courts. Here is where AI in legal reasoning can be of great benefit.

A general legal ES has a rule-based structure. Its purpose is to provide expert legal advice about such specialized areas as:

- Pension eligibility.
- Social security eligibility.
- Licensing requirements.

These types of expert systems can be created where the law is relatively complete and independent but still complex enough to require specialized expertise—areas where the law is relatively static, not changing rapidly. In these areas, it is possible to compartmentalize the law into small chunks for coding into the knowledge base by asking users simple questions, the answers to which generate the ES response. Then it is only necessary to update the ES knowledge base with current law.

AI developments in modeling case-based legal reasoning have generally focused on predicting the outcome of litigation by assigning weights to case facts. The accuracy of the prediction depends on how similar the case being currently litigated is to previous cases that have been decided. These expert systems rely on reasoning by example by considering the precedent. Rules derived from previous cases are used to predict the outcome of current cases. Such expert systems are domain-specific and limited to specialized areas or topics. They are gradually being used for decision-making purposes, such as for judicial sentencing.

FINDER

FINDER is an example of a predictive case-based expert system. Its domain of knowledge consists of cases involving disputes about property and property rights. A typical case involves a dispute between an individual who finds something on real estate occupied by another party. The user is asked a set of "yes" or "no" questions, and the answers elicit a prediction.

The way it works is this: After the questioning is completed, FINDER enters its computational mode. The facts entered contain the information needed for the analysis of case law. The algorithm selects a predicted outcome to the case on the basis of the closest precedent case in the system's knowledge base. This case is provided as the authority for the predicted outcome. The algorithm then searches through its knowledge base for the case that is closest to the client case that has the opposite outcome.

The written opinion first specifies the predicted outcome and cites the most relevant authority, with a description of its most important similarities with the client's case. The algorithm then repeats the process for the distinguishing case, noting its most important differences from the client's case.

To confirm the prediction, the FINDER uses an alternative method. If this does not yield consistent results, FINDER warns the user that the case is "difficult" and should be analyzed by a human lawyer.[1]

SHYSTER

The FINDER system, along with three other specifications of Australian law, is included in a case-based expert system known as *SHYSTER*.[2]

To test SHYSTER and its approach to case law, four specifications were written. Each specification represents an area of Australian law:

1. *Finder specification.* An aspect of the law of property and property rights.

2. *Authorization specification.* The meaning of "authorization" in copyright law.

3. *Employee specification.* The categorization of employment contracts.

4. *Natural specification.* The implication of natural justice in administrative decision making.

SHYSTER produces several intermediate files in the process of creating its reports. The following are produced each time SHYSTER is invoked:

- Operation and warnings messages are summarized in the *log* file.

- Internal representation of the case law specification is contained in the *dump* file.

---

[1] For more information about FINDER see *http://www.law.usyd.edu.au/~alant/ai.htm* and *http://www.law.usyd.edu.au/~alant/aulsa85.html.*
[2] For more detailed information about SHYSTER see *http://cs.anu.edu.au/software/shyster.*

- Dependence probability figures for each attribute pair in each area of law in the specification are contained in the *probabilities* file.
- A *weights* file gives details of the weights assigned to each attribute.
- A *distances* file provides similarity measures for the instant case.
- A *report* file is produced for each invoked area.

The following twelve modules make up the SHYSTER system:

1. *Shyster* is the top-level module for the whole system.
2. *Statutes* is the top-level module for a rule-based system.
3. *Cases* is the top-level module for the case-based system.
4. *Tokenizer* tokenizes a program written in SHYSTER's case law specification language.
5. *Parser* parses a program written in SHYSTER's case law specification language, using the tokens.
6. *Dumper* displays the information that has been parsed.
7. *Checker* checks for evidence of dependence between attributes.
8. *Scales* determines the weight of each attribute.
9. *Adjuster* allows the legal expert to adjust the weights of the attributes.
10. *Consultant* interrogates the user as to the attribute values in the particular case.
11. *Odometer* determines the distances between the leading cases and the case at hand.
12. *Reporter* writes SHYSTER's legal opinion.[3]

APPLIED SENTENCING SYSTEM

Applied Sentencing System (ASSYST) is an example of a case-based decision-support legal ES that was developed by the U.S. Sentencing

---

[3]For sample outputs from SHYSTER see *http://cs.anu.edu.au/software/shyster/output.*

Commission. It helps determine the guideline sentencing range using a knowledge base of statutes and cases.

LEGAL INFORMATION RETRIEVAL SYSTEM

Legal Information Retrieval System (LIRS) is a case-based ES on negotiable instruments law (from Uniform Commercial Code, Articles 3 and 4). Carole D. Hafner developed it. LIRS used a semantic network model of approximately 300 concept nodes. Its knowledge base contains approximately:

- 186 cases.
- 110 sections of the UCC statutes.
- 188 official comments.

THE NERVOUS SHOCK ADVISOR

The *Nervous Shock Advisor* (NSA) provides advice about damages for the negligent infliction of nervous shock, otherwise known as "post-traumatic stress disorder." It was created at The University of British Columbia Faculty of Law Artificial Intelligence Research Laboratory (FLAIR) using a commercial expert system shell. The topic of nervous shock has well-defined limits within the broader context of the law of negligence. It refers to claims for shock caused by a traumatic event similar to the action of *emotional distress*. For example, if a bystander witnesses something horrible and suffers shock (emotional distress) as a result, courts may in certain limited circumstances award damages against the person who caused the shock. The aim of this expert system is to predict the outcome of cases where otherwise the outcome is unclear.

NSA asks the user questions about a case to elicit facts to see if they satisfy all the material elements required. If so, the NSA conclusions are supported by precedent cases that match the facts provided by the user. If not, if the outcome is predicted to be negative, reasons are provided and precedent cases are presented to justify the conclusion. The appendix to this chapter demonstrates how NSA works in a typical case.

NEGLIGENCE

An experimental negligence expert system is currently available on the Web. Its aim is to help lawyers determine whether there has been negligence in a particular situation.

DOMESTIC VIOLENCE

An experimental domestic violence expert system is available on the Web. Its system is designed to model options available in cases of domestic violence.

ICAR

ICaR is an expert system based on the Intellectual Information System. ICar:

- *Consults* in applying captured expert knowledge to generate an opinion on a specific case described to it by the user.
- *Explains* an expert's opinion and lets the user run "what-if" simulations and "how-to" plan consultations.

Virtual Tax Adviser is a shareware application based on an ICaR expert system shell. It may be used for consulting and reasoning on United States tax questions.

THE INHERITANCE ADVISOR

The *Inheritance Advisor* (IA) is designed to help lawyers determine the proper distribution of assets for a person who has died intestate. The system helps the lawyer collect all the necessary information.[4]

IA uses a backward chaining procedure with a depth-first search. While the laws of intestate distribution are similar in each state, they are nonetheless state-specific. IA has the intestate distribution laws of the state of Oklahoma in its knowledge base. The statutes were interpreted by practicing lawyers on the development team.

---

[4]*www.isworld.org/ais.ac.98/proceedings/track04/blackstock.pdf.*

Legal Applications

The IA ES conducts an interview to systematically collect information to make sure no important facts are overlooked. The system then can quickly and consistently apply the law in its knowledge base to the facts so as to provide recommendations immediately after the interview ends.

APPENDIX 6-A: SAMPLE CONSULTATION HYPOTHETICAL FOR NSA[5]

A man is standing on a downtown street talking to a friend. His 10-year-old son is further down the same block on the sidewalk, watching workers on a construction site where a new office tower is being built. Though the skeleton of the building is in place, work is going on at several different levels that have not been enclosed. On one of the upper floors, a workman has failed to secure a piece of machinery properly. The machine begins to roll toward the edge of the building where it borders the street. One of the workers yells a warning just before the machine rolls over the edge. The father turns and sees the machine fall and crush his son to death. He suffers nervous shock. He is often depressed about his son's death and sometimes contemplates suicide. Can he recover damages for the shock?

Consultation

```
FLEX >go
*-*-*-* Welcome to Nervous Shock Advisor *-*-*-*

I will tell you whether or not your client has a
cause of action in nervous shock. Simply type
your answers to my questions on the keyboard
located below the screen. If you wish to know why
a particular question is being asked, feel free
to type "why" in response to the question. You
may also type "unknown" if you are unable to
answer a particular question. If so, I shall can-
vass whatever other options are available to me.
```

(The banner appears whenever a consultation is started, but will not be duplicated in the subsequent examples.)

```
Was the incident that caused the nervous shock
the result of an act on the part of the defen-
```

---

[5]*Source:* C. Deedman and J. C. Smith, "The Nervous Shock Advisor: A Legal Expert System in Case Based Law," *Operational Expert Systems in Canada,* (Pergamon Press: Elmsford, NY: 1991).

```
dant that you, as a lawyer, would characterize
as negligent or deliberate with respect to any
of its harmful results?
1. negligent
2. deliberate
        >>1

Was the nervous shock sustained as the result of
an incident in which any person or persons,
including the plaintiff, suffered physical harm
or the risk of such harm?
1. harm
2. risk
3. both
4. neither
        >> 1 ~
```

(*Note:* Option number three necessarily implies that there was more than one victim.)

```
Who was the victim of the incident that caused
the nervous shock: the plaintiff, a third party,
or both?
1. plaintiff
2. third party
3. both
        >>2
```

(*Note:* If there was more than one third-party victim, your responses to any questions about a third party should relate to the most serious case only.)

```
Was the other person killed?
>>yes

Is your client closely related to the other per-
son?
>>yes
```

```
Is the other person a member of your client's
immediate family; namely, the child, spouse,
parent, or sibling of your client?
1. child
2. spouse
3. parent
4. sibling
5. neither
      >>1
```

(*Note:* Regard a step-relationship as equivalent to a natural one if the person was treated like a member of the natural family and a grandparent as equivalent to a parent if the grandparent had been filling the role of a parent.)

```
Did your client see the incident happen?
>>yes

Did the nervous shock cause an expense, a loss
of income, or an incapacity that was sufficient-
ly serious to support a claim for damages?
>>yes

Into which of the following general categories
did the symptoms of nervous shock fall: physical
illness, mental disorder, sleep disturbance?
   Separate each option by a comma if more than
one is applicable.
1. physical illness
2. mental disorder
3. sleep disturbance
      >>2

JUDGMENT: An action lies 100%.

The following is a breakdown of the elements
involved in this decision. If any element has a
certainty factor of less than 100%, you should
regard it as a weak link in the chain and bol-
ster your case in that area.
```

Negligence established 100%. Appropriate triggering event 100%. Third-party principal victim 100%. Certainty not increased because only one victim involved. Sufficiently serious incident 100%. Plaintiff connected to the victim 100%. Plaintiff exposed to incident 100%. Damage proved 100%.

I am pleased to inform you that your client has a cause of action for nervous shock.

Would you like to peruse the cases that support my decision?
>>yes

Press Enter to continue . . .

THE FOLLOWING CASES ARE RELEVANT TO YOUR FACT SITUATION:

DEATH--CHILD--WITNESS CASES ON POINT:
1 Howes v. Crosby (1984), 45 O.R. (2d) 449, 6 D.L.R. (4th) 698, 29 C.C.L.T. 60 (C.A.).
2 Richards v. Baker, [1943] S.A.S.R. 245 (S.C.)
Other Family Relationships
3 Storm v. Geeves, [1965] Tas. S.R. 252 (S.C.).
4 Walker v. Broadfoot, [1958] O.W.N. 173 (H.C.).

RELEVANT BY ANALOGY:
Injury--Related--Witness
5 Pollard v. Makarchuk (1958), 26 W.W.R. 22 (Alta. S.C.).
Pg Dn to continue . . .
Enter the number of the case you wish to see:
>>1

(*Note:* When the selection is made and the Enter key pressed, the text of the case summary appears on a new screen.)

Howes v. Crosby (1984), 45 O.R. (2d) 449, 6 D.L.R. (4th) 698, 29 C.C.L.T. 60 (C.A.).

FACTS
The plaintiff's mother was walking along the shoulder of a road holding her young daughter's hand when a car driven negligently by the defendant struck the girl from behind. The brother and sister of the victim were also walking with their mother and witnessed the accident. The father came to the scene immediately and attempted to give mouth-to-mouth resuscitation to his child but the young victim died. The entire family sued and recovered for nervous shock.

SYMPTOMS
P. 452 . . . "As noted earlier, the attack is made on the awards to Susan and David. Janet was 5 at the time of her death, Susan was 7 and David was 16.
(a) Susan

(*Note:* In cases that are relevant for their FACTS, the full SYMPTOMS portion of the case will not be shown.)

[Exit from data file.]
You said that your client had mental problems. Would you like to peruse the cases that relate to these symptoms?
>>yes

Press Enter to continue . .

THE FOLLOWING CASES ARE RELEVANT TO MENTAL SYMPTOMS:
1 Walker v. Broadfoot, [1958] O.W.N. 173 (H.C.) (depression, suicide).
2 Hinz v. Berry, [1970] 2 Q.B. 40, [1970] 2 W.L.R. 684, [1970] 1 All E.R. 1074 (C.A.) (depression, suicidal ruminations).
3 Brice v. Brown, [1984] 1 All E.R. 997 (Q.B.) (insanity).
4 Tsanaktsidis v. Oulianoff (1980), 24 S.A.S.R. 500 (S.C.) (severe depression).

5 Montgomery v. Murphy (1982), 37 O.R. (2d) 631, 136 D.L.R. (3d) 525 (H.C.) (depression).
6 Kohn v. State Government Insurance Commission (1976), 15 S.A.S.R. 255 (S.C.) (intensified depression).
7 McLoughlin v. O'Brian, [1982] 2 W.L.R. 982, [1982] 2 All E.R. 298 (H.L.) (personality change).
8 Fenn v. City of Peterborough (1976), 14 O.R. (2d) 137, I C.C.L.T. 90, 73 D.L.R. (3d) 177, varied
25 O.R. (2d) 399, 9 C.C.L.T. I (C.A.) (loss of pride and self-respect).
9 Hogan v. City of Regina (sub. nom. McNally v. City of Regina), [1924] 2 W.W.R. 307, 18 Sask L.R. 423, [1924] 2 D.L.R. 1211 (C.A.) (amnesia).

(*Note:* The interaction with NSA continues till the user wishes to stop. NSA continues to provide additional detailed information.)

CHAPTER 7

Artificial Intelligence in Marketing

To fully appreciate the extent to which marketing as a business activity has embraced artificial intelligence, it's necessary to take a step back to examine the purpose of marketing as a business activity. One of the easier ways to examine this is to look at the definition of marketing.

Different scholars and businesspeople define marketing in different ways. You can find a good sample in Philip Kotler's marketing management text.[1] For example, he quotes Peter Drucker, one of the gurus of management, as having defined marketing as "so basic that it cannot be considered a separate function. It is the whole business seen from the point of view of its final result, that is, from the customer's point of view." Ray Corey said marketing consists "of all activities by which a company adapts itself to its environment—creatively and profitably."[2]

One thing these two definitions have in common is the recognition that marketing comprises components on which the very existence of the organization hinges. Marketing activities can be divided into four main functions:

---

[1] Kotler, Philip. *Marketing Management: The Millennium Edition.* Englewood Cliffs, New Jersey: Prentice-Hall, 2000.
[2] *Ibid.,* pp. 7-9.

1. Helping design the product or service.
2. Selling.
3. Forecasting.
4. Communicating.

We will examine briefly how AI can work with these marketing functions to:

- Enhance the quality of work.
- Execute tasks more easily.
- Improve efficiency and productivity.

Today's marketing manager makes numerous decisions that have a major impact on the profitability of the firm. For example, in many firms one makes a series of what is technically referred to as "go, no/go" decisions when new products are being designed and produced. A "go" decision moves the innovation process to the next level, and a series of "go decisions" will eventually end up with commercialization of the product. A "no go" decision in the innovation process at any point essentially kills the project.

What is the likely impact if a decision is made to launch a product that nobody wants, or abort a project later successfully marketed by a rival company? What about the likely impact of producing the wrong quantity of a product? Is it better to be cautious and market only a small quantity? If there's high demand, there will be a lost sales opportunity. On the other hand, if a company produces more than the market will bear, money will be tied up in excess inventory. Managers often have to make such far-reaching decisions with less than perfect information, under a huge cloud of uncertainty.

That is why marketing has embraced artificial intelligence. Marketing departments that adopt AI in some form are better equipped to deal with the new external environment within which they have to conduct their activities.

An AI-equipped marketing department is also better equipped to interface with other firms that use AI to process their orders or otherwise to conduct business. For example, without adopting the technology firm

A might not be able to serve the needs of firm B, a client that uses the technology to share information with its clients.

Do workers in marketing departments have the skills they need to successfully adapt, develop, and operate AI software? Or to reverse the question, are AI personnel from other departments welcomed in marketing departments? Using AI successfully in marketing cannot be made up of separate entities. It calls for continuous innovation. That means it must involve personnel from both operations research/product management and marketing departments. Organizations that are successful in integrating AI into marketing realize that the success of the entire organization is more important than inter-department rivalry. They are carefully structured for successful interface between the two departments.

Marketing's embrace of AI also depends on the state of the arts of AI itself—what it does and can do to enhance marketing. Can AI help reduce errors in marketing decisions? Many good products or services have failed in the marketplace because of lack of demand. John DeLorean's gull-winged stainless steel sports car was technically sound, but not only did it fail, so did the company.[3] Kodak's much-touted *Advanta* camera also failed to score a hit with consumers, further depressing the company's dwindling advantage in the photographic equipment market. For a product to be successful, it must have the features consumers want at the price they will pay. How can marketing managers know what features consumers want?

USING AI AT THE DESIGN STAGE

Because the new product innovation process is very expensive and yet the failure rate of new products is very high, companies actively seek ways to reduce the chances of failure. Marketing managers solicit input from potential consumers in a variety of ways, most of which we will discuss under marketing research and forecasting. Here we will just note that an increasing number of marketing managers enlist the help of AI, directly or indirectly, when they design new products.

One common way to reduce the chance of failure is to try and determine the utility value (in simple terms the level of satisfaction) the con-

---

[3]For a brief review on the life of the DeLorean sports car, see Morrison, Ann M., "Dream Maker: The Rise and Fall of John Z. DeLorean," *Fortune* (Sept. 19, 1983).

sumer attaches to different product features and thus ultimately to the product itself. The higher the perceived utility value, the more desirable the product should be, so features are added or eliminated based on their utility value.

A simple comparison of three toothpastes with three different packaging styles, three different brand names, and three different price levels will produce about 27 different combinations (three times three times three), already a fairly large number for anyone to simultaneously consider. That's before we even introduce attributes like teeth-whitening, and cavity-preventing abilities that a potential toothpaste buyer is likely to care about. Because of the serious limitations inherent in direct comparisons, marketing managers use a statistical technique known as conjoint analysis in assessing utility value. It involves asking respondents to rank different hypothetical products whose attribute levels are deliberately changed. The ranking allows a producer to determine whether product A would be preferred to product B and product B to product C. The usefulness of the conjoint technique goes beyond the relative ranking of products, because it also allows a producer to determine the relative importance of each attribute.[4]

We must also emphasize that the use of the conjoint technique is not limited to tangible products like toothpastes, cars, or microwaves. Competitive firms in the hotel, airlines, and financial services industries—in fact, even academic institutions—use conjoint analysis to help decide what new features to incorporate in their offerings.

How does AI relate to conjoint analysis? While it's perfectly possible to conduct conjoint analysis without using AI, the complexity and volume of data collected by marketing researchers these days are constantly testing the computational limits of traditional statistical software packages. One of the advantages of AI, particularly neural networks, is its ability to analyze large data, learn complex patterns, and abstract generalized information very quickly.

It is evident from the toothpaste example that with AI, the number of products and attributes considered can be greatly increased. Furthermore,

---

[4]For a more detailed discussion of conjoint analysis, see Malhotra, Naresh K. *Marketing Research: An Applied Orientation,* 3d. ed. Englewood Cliffs, New Jersey: Prentice Hall, 1999. For an interesting but technical application of the technique, see Carroll, J. Douglass and Green, Paul E. "Psychometric Methods in Marketing Research: Part I, Conjoint Analysis," *Journal of Marketing Research,* 32 (Nov., 1995): 385-91.

as a statistical technique, conjoint analysis produces only an estimate, which can easily be off target. Combining conjoint analysis with AI reduces the error level significantly, greatly improving the accuracy of estimates.

SELLING

Selling is one marketing area into which AI has made a foray. As you have already seen earlier in this book, AI is a subset of computer science. It deals primarily with expert systems, natural language, and neural networks. Marketing activities that call for voice and pattern recognition are easily amenable to AI. And while inventory control could be considered the back-end of the selling aspect of marketing, depending heavily on computer applications, the front-end AI application would clearly be the virtual store.

In its most basic form, a virtual store is an electronic catalog maintained at a portal that can be accessed by any potential customer with access to a computer and the Internet.

The Internet

A basic Web site simply lists the items that the company is offering to the public, takes orders, and calculates the total expenditure. It is also used to collect information on the potential buyer. Shopping in a virtual store can be pleasant or frustrating, depending on how easily the customer can navigate the site.

Incorporating AI programs into Web sites has made navigating much more pleasant. For example, Pillsbury was getting so much demand for time-saving cooking tips on its Web sites that in 1997 the company implemented a Web site architecture that could identify and tailor responses to each visitor based on that visitor's Web browser. The program intelligently anticipates what information the user wants and provides an accurate answer. The time savings that the customer enjoys should lead not only to increased sales but also to an increase in goodwill towards the company.

Another example of a clever use of AI in Web site design is illustrated by cozone.com Inc.'s use of Jill, a CRM (customer relationship management) program that helps customers find what they're looking for. Jill uses predictive modeling, demographics, and complex algorithms to assess customer needs, but it asks customers questions in plain English, evaluates their responses, and suggests a product that it "thinks" is appropriate.

It's important to keep in mind that Web sites are not always directed towards final consumers alone. They're regularly used in business-to-business (B2B) marketing as well. For example, General Electric, one of the world's largest business "consumers," now uses the Internet to buy most of its operating supplies. Where does GE buy them? Naturally, from other businesses.

Other examples of companies doing B2B marketing via the Internet are Japan Airlines and National Semiconductor. Even governments do it. Los Angeles County reportedly buys an enormous amount of goods and services over the Internet.

Efficiency and cost cutting are the major reasons for motivations behind B2B transactions over the Internet. With the increasing intensity of competition, companies are perennially searching for ways to cut costs. B2B Web selling helps companies reduce both turnover time for orders and transaction costs.

Many companies also reduce mistakes in ordering by incorporating some aspect of AI into their B2B Web sites. For example, pattern recognition programs can detect when a buyer mistakenly places an order that is a departure from its routine purchases. Without AI programs such mistakes go undetected and result in the wrong order being filled, only to be returned, at a cost in time and money for both parties. Pattern recognition programs nip mistakes like this in the bud—when a prompt statement alerts the operator to a possible error, the order is verified. It can still go through if the departure from past patterns is intentional, but if it's not, time-consuming and sometimes expensive delays are avoided.

Because of its enormous potential, there's a constant search for more ways of applying AI in B2B marketing. Not long ago, Fish, Barnes, and Aiken explored how neural networks (NNs) might help B2B marketers faced with market segmentation problems.[5] Applying discriminant analysis and logistic regression traditionally does the job, but the researchers achieved higher hit ratios on holdout samples using NNs. Thus, a mar-

---

[1]Fish, K.E., Barnes, J.H., and Aiken, M.W. "Artificial Neural Networks: A New Methodology for Industrial Market Segmentation," *Industrial Marketing Management* 24 (1995). 431-39.

keter pitching to industry may be able to segment the market more accurately with NNs and thus more efficiently deploy its sales force and design more effective promotions.

Direct Marketing

Now let's turn to another aspect of selling activities in marketing—direct marketing. Direct marketing is a strategy that allows the marketer to bypass the middleman and make direct contact with the target market.

The Final Consumer

Direct marketing to the final consumer dates back to the agrarian economy when the farmer/producer sold directly to the person who would cook and eat the food, but it has undergone major changes since new technologies have been incorporated into business practices. The legendary Sears catalogs, the now popular catalogs from Lands' End, and Avon's door-to-door salespersons embody the concept.

The problem is that this type of direct selling is not efficient. Endless number of catalogs get tossed into the garbage bin because they were sent to the wrong targets. Endless number of doors are shut in the faces of door-to-door salespeople who have knocked on the wrong door. These wastes could be eliminated if catalogs are sent to the right targets—people who need and will use them, who will consider the catalog a service instead of a nuisance. Fewer doors will be shut in the face of the salesperson who calls on the right homes, where the call will be considered a service instead of an interruption.

Incorporating AI into direct marketing is an attempt to solve the problem of wrong targeting. The source of this problem, of course, is the dearth of the information on individuals needed to hit the right target. Because they can analyze huge amounts of data, AI-based programs allows marketers to access this information usefully. The technology also gives marketers ways to get information they didn't have before. "Cookies," the technology used to collect information on a Web browser's habits, are primarily controversial because they work.

Armed with this new information and supported by the advanced analytical capacity provided by other AI-based programs, marketers can now create more accurate consumer profiles and as a result more accurately select persons who are more likely to buy from them. For instance, a stockbroker soliciting new business by mail can now segment target markets on the basis of income, lifestyle, hobbies, and outlook on life, and forget about the old "average investor."

An example of a program that enhances segmentation and targeting ability is PRIZM (Potential Rating Index by Zip Markets) sold by Claritas, Inc. PRIZM is a geoclustering technique that allows the formation of 62 distinct clusters or lifestyle groupings of over half a million residential neighborhoods. The clusters are based on as many as 39 factors in five broad categories (1) education, (2) family cycle, (3) urbanization, (4) race, and (5) mobility.[6]

Among other benefits, pattern recognition technology allows marketers to segment customers on the basis of purchase patterns, so they can give these customers advance notice of the arrival of new products that those customers are likely to find most useful. Say analysis of the data reveals that one consumer, Jose, buys a mystery book every five weeks or so. Jose seems to be a more likely buyer for a new mystery soon to be released than is Peter, who often buys from the same online company but buys wine, not books. Jose gets the promotional material and Peter does not.

Thus, pattern recognition technology allows the marketer to serve Jose as though he belongs to a book club without his having to deal with membership applications or dues. The same marketer can use the technology to let Peter know of new wines. Jane, who buys both books and wine regularly, will be informed about both.

The bottom line is that the technology makes the marketer more efficient. It reduces the clutter of information in people's mailboxes (both electronic and snail mail). It reduces waste. Because better targeting means reaching only people who are more likely to buy, productivity increases. The final consumer benefits from the technology because the cost savings enjoyed by the marketer are passed on as lower prices.

Wine and books are small-ticket items, but the technology works just as well with big-ticket items. Car manufacturers do the same type of things, though they're more likely to collect their data from question-

---

[6]For more information on PRIZM, see Del Valle, Christina. "They Know Where You Live—and How You Buy," *Business Week* (Feb. 7, 1994). For a case study on PRIZM see Englis, B.G. and Solomon, M. "To Be and Not to Be: Lifestyle Imagery, Reference Groups, and the Clustering of America," *Journal of Advertising*, 24, 1 (1995): 13-29.

naires filled out or from cookies at Web sites and not from online purchases.

AI has significantly affected other aspects of direct marketing as well. For example, speech recognition technology is now being used in telemarketing. Devices that can hold simple conversations are used to dial and talk to potential customers, to screen calls, or to prequalify potential customers. Eventually, they might be used to collect data. Fenn and Hodgon predicted in 1995 that virtual reality would soon allow marketers "who currently rely on video presentations to add another dimension to product viewing."[7]

Marketers can even apply AI technology in "showrooms." Because it's now possible to portray objects vividly in three-dimensional space and even rotate the object so that it can be viewed from different angles, even manufacturers that don't actually sell online, like Jaguar or Mercedes, maintain Web sites that are like real life showrooms, displaying their offerings and instructing "visitors" on how to get in contact with the nearest dealer. Because these sites give potential buyers valuable information, it decreases the time they later have to spend with a salesperson during the actual buying process.

The Business Consumer

Businesses typically buy things in order to further a production process. A small business account with Dell may be completely different from a household account. The small business might be buying a computer to keep track of its customers, manage payables, produce invoices for receivables, and do heavy volumes of word processing. A household might be buying it to gain access to the Internet, keep track of a small number of bills, and do minimal word processing. Because of their different computing needs, household and business computers may have to be configured differently.

Another distinction is that while a household might buy one or two computers, businesses might buy dozens or hundreds and might need different kinds of service arrangements. The sales approach has to take all that into account.

---

[7]Fenn, J. and Hodgdon, P. "How Emerging Artificial Intelligence Technologies Will Affect Direct Marketing," *Direct Marketing,* 58, 4 (1995): 28-31.

Selling directly to business customers instead of through middlemen allows manufacturers to observe directly how their clients operate, the steps the purchasing process goes through, and how products are used. With the cooperation of their clients they can collect high-quality data from such observations. By analyzing these data using AI-enhanced statistical techniques, most manufacturers can thus offer their customers suggestions on how to save, alternative ways to produce products, or develop a new and cheaper product that should satisfy the same objectives.

With the right technology and the cooperation of their business customers, producers can collect information not just once but continually. They can analyze the data regularly to determine whether clients may have new needs and to advise clients of what might be useful to them.

Such an analysis in the computer industry recognized that downtime is very costly to business clients. To mitigate these costs, computer companies like Dell and Tandem use AI-enhanced technology for repairing products remotely on the premises of their clients—even before their business customers become aware that there is a problem. Using remote diagnostics they can detect problems and respond before the problems lead to costly downtime.

Most of the many presentations involved in B2B take place where the client is located, so traveling constitutes a large part of the expenses associated with selling. In addition to the direct costs of transportation, accommodation, and meals, these trips also tie up the sales team's time. When presentations can be done remotely without the presenters being physically onsite. The cost savings can be enormous. Such presentations were in the past hampered by poor-quality transmission, but AI-enhanced technology has made high-quality presentation and two-way communication possible, thus reducing the cost of doing business.

Trade shows have been a traditional showcase where firms present their current offerings and highlight new technologies and future products, but like on-site presentations, trade show attendance is very expensive. There are all the usual traveling costs for personnel plus additional costs for transporting sizeable amounts of inventory. With advanced graphics capabilities, however, considerable attention has shifted back to virtual showrooms, so attendance at trade shows is no longer essential for reaching new customers.

Thanks to AI applications, teleconferencing technology has now improved enough that pictures of participants can be seen and their voices heard without delays and echoes. This has greatly facilitated not only B2B marketing but also intracompany meetings across different geographic locations. Because innovations like this lead into more efficient ways of doing business, the ultimate consumer invariably benefits from the cost savings.

MARKETING SERVICES

An ever-increasing percentage of the American economy is service based, and AI has been effective here, too, in enhancing marketing for the intangible offerings of banks, hotels, schools, fire stations, hospitals, and the telecommunication industry.

Because competition in delivering services is as intense as competition in delivering tangible goods, producers of services are also engaged in a constant search for ways to improve both their offerings and their delivery. For example, in an effort both to maintain its reputation as the industry's top innovator and to improve its market share, AT&T's Universal Card Services invested heavily in an AI marketing system that uses NNs, pattern recognition programs, and fuzzy logic to more accurately identify a larger group of potential credit card users. Because AI-based programs more closely mimic human reasoning, they out perform AT&T's traditional statistical programs for identifying target markets.

Financial markets are far from immune to competition. Since the industry was deregulated, competition for consumers has intensified greatly, so financial institutions are devoting significant resources to developing expert systems that formulate marketing strategies. *BANKSTRAT*[8] is a program used by retail banks and building societies. Once the program has been given information on market size, types of banking products, profitability, marketing expenses, and the level of competition, it recommends strategies on expanding geographically and on market penetration.

Like financial institutions, hospitals and hotel chains are getting into the AI act. Even nonprofit organizations are reportedly using AI programs

---

[8]For a detailed discussion on how BANKSTRAT works, see Davies, F., Moutinho, L., and Curry, B. "Construction and Testing of a Knowledge-Based System in Retail Bank Marketing," *International Journal of Bank Marketing,* 13, 2 (Feb.) 1995: 4-15.

in fundraising. For instance, NN modeling is used to predict the behavior of potential donors, current donors, and supporters. Because human behavior is complex, AI models, themselves complex, are used to predict human behavior. But a well-structured NN can "effectively assume a set of rules that can predict the individual donor's responsiveness for future donor acquisition campaigns."[9]

FORECASTING

Here we use the word forecasting in a broad sense to mean all activities in which something is estimated. That makes it clearer why forecasting plays such a preeminent role in marketing.

Marketing basically finds what consumers need or want and provides something to fulfill the need or want. Yet many consumers don't themselves know exactly what they need or want. Sometimes they do know vaguely what they need but aren't able to clearly articulate it. To fulfill unmet needs or wants, a good marketer has to be careful not to take at face value all the answers consumers give to the marketer's questions.

This, of course, suggests that success in forecasting consumer demand lies in first figuring out exactly what it is the consumer wants or needs. That is why our discussion of AI applications in forecasting will cover consumer behavior (how the marketer figures out the consumer) and market research (how marketers systematically gather and analyze data). We will consider these in the context of forecasting, inventory management, and trend analysis.

Consumer Behavior

In order to design and skew their promotional strategies toward the most likely decision, marketers want to understand what motivates the consumer to make a purchase and how the consumer makes a decision when confronted with alternatives.

Their attempts to understand run into two main problems. First, they have to work with patterns (models) that are either unstructured or hard to identify. Secondly, they have to work with a large number of qualitative variables. For example, a consumer's purchase decisions appear to be

---

[9]Goodman, S. and Plouff, G. "Neural Network Modeling: Artificial Intelligence in Marketing Hits Non-Profit World," *Fund Raising Management,* 28, 4 (1997): 16-18.

motivated by at least four major factors (1) cultural, (2) social, (3) personal, and (4) psychological. Each factor has several subfactors. The cultural factor, for instance, includes both the family and institutions through which individuals acquire their values. Additional subfactors might be religion, race, or geographic region of residence.

How these factors interact in a purchase decision are complex. Traditionally, marketing researchers use statistical techniques known as choice models to predict a consumer's likely selection from given alternatives.[10]

Standard statistical techniques are limited in their ability to deal synchronously with a vast array of qualitative data types. The problem is compounded here by the sheer volume of data to be analyzed. The result is predictions that are less than accurate. Fortunately, NNs have arrived to take up the slack. They can learn a wide range of complex patterns of precisely the kind that occur when consumers make purchase decisions.

To find out how consumers evaluate their options, market researchers now use a combination of NNs and statistical modeling. Together they enable the system to mimic actual human decision making behavior using its rule-induction capability. They not only predict outcomes better, they also shed light on how an actual human being would have made a given decision. Insights like that can give a firm a real competitive advantage.

Gathering the large amounts of data needed by continually monitoring the consumer would be time-consuming, expensive, and ultimately not practical. However, advanced remote sensing technology can, with the consent of the consumer, gather the information through inconspicuous monitors placed at remote locations to observe the consumer.

The Actual Forecast

The marketer, now armed with knowledge of the consumer's needs or wants, must at this point forecast sales or predict market share. No one wants to run short of a product that is in demand ("moving fast"). No one, on the other hand, wants to be saddled with excess inventory when the offering is not in demand (not moving fast). Stock outs on fast-moving items not only lose the sale but could incur loss of customer goodwill.

---

[10]For a detailed discussion on the theory and application of the choice model, see Ben-Akiva, M. and Lerman, Steve R. *Discrete Choice Analysis: Theory and Application to Travel Demand,* 3d. ed. Cambridge, Ma: MIT Press, 1989.

Overproduction entails lost opportunity, cost of funds, and the expense of carrying excess inventory. "Overage" and "shortage" are dreaded words in marketer discourse.

To solve the production dilemma marketers engage sophisticated statistical techniques to estimate future demand levels on the basis of past demand levels. Different regression techniques are popular, but they have a common weight estimation problem that can lead to either over-fitting or under-fitting, and thus to a serious inaccuracy in the forecast. Furthermore, regression techniques cannot generally produce reliable forecasts that fall outside the range of the data being used.

Incorporating AI techniques into forecasting has helped to overcome these two problems. Take NNs, which can learn complex patterns, for example. Using NNs in market response modeling, by eliminating the need to estimate weights, has notably reduced over and under fitting problems. Even though an NN has to be "trained," it outperforms regression techniques on holdout-sample data.

Predicting market share is another important aspect of marketing that relies heavily on forecasting techniques. Decisions on whether or not to launch a new product or service are made partly on the basis of the projected market share that the product or service will most likely take. It's no surprise, therefore, that market share projections are essential for making profit and loss projections. They are also important aspects of the firm's strategy. We will talk more about this later.

To highlight the extent to which forecasting has benefited by integrating AI, let's run through the activities involved in forecasting (though there are no hard and fast rules for delineating these activities). There are four main steps:

1. Theory development.

2. Data collection.

3. Data analysis.

4. Output generation.

Theory Development

Here marketers create a theory that captures the activity or activities of interest and identifies a feasible relationship among factors (this process is also referred to *developing a construct*). It's important to have a theory because marketers need to be able to generalize the model for similar situations.

Data Collection

The information gathered from a given situation at hand could be in verbal or numeric form.[11] Generally, for instance, qualitative data in regression models is captured by dummy variables so everything is numeric. This is the input material from which the model is created. Any mistakes made in data collection can affect the accuracy or the functionality of the model. (Though there is much that can be done in the data collection process to ensure the integrity of the data, further discussion is outside the purpose of this book.)

Data Analysis

In the analysis phase the information gathered and the model developed are linked through intensive application of a variety of analytical techniques. Usually, today, analysis requires a combination of statistical and AI-based programs. The choice of technique depends on not only the complexity of the data but also on such other things as the type of data collected—verbal or numeric and if numeric, whether they are nominal or ratio.

Output Generation

The end product of the process is a model—in our context, the forecast. AI impacts sales or market share forecasting in several ways, particularly in the two previous steps. It helps collect data that could not have been collected through the traditional means, such as the information collected by cookies or through observations at remote locations.

While it would be preferable that all data used for modeling be actually collected ("real"), sometimes access to real data is for one reason or the other limited. For example, when a product or service provider prefers not to actually test the market in an entire geographic location or on the complete target market, data based on a small sample of respondents are

---

[11]Some authors distinguish between information and data, preferring to use information to refer to interpretation of data. We use them interchangeably.

used in the simulation models that predict market share or forecast sales volume.

The accuracy of NN-based simulation models, despite their computational complexities, has won the admiration of those who use these types of predictions to make important marketing decisions. Widely used simulation models are *COMP,* by Elrick and Lavidge, *ASSESSOR* by Management Decisions Systems, Inc., and *SPEEDMARK* by Robins Associates.

Though it is often discussed as an afterthought, presentation of the outputted model is an important aspect of marketing activities, especially to marketers and consultants who have to present their findings to clients. Even here the usefulness of AI-based programs cannot be overlooked. The same technologies we discussed earlier in relation to showrooms and trade fairs can be effectively adapted to presentations for clients.

INVENTORY MANAGEMENT

Though inventory management typically falls into the domain of an operating department, the recent trend towards a crossdisciplinary approach to managing core processes has more intimately involved the marketing department in it. The central concern of inventory management is to determine the optimal combination of inputs and outputs. In other words, it seeks to prevent excess inventory of finished goods by controlling output and preventing excess inventories of raw materials by controlling purchases.

Although we will not discuss how AI programs are deployed generally in operations, we will talk about how marketing departments have brought to bear on inventory management the AI-based techniques that are integrated into marketing.

Some companies have developed the art or science of inventory management to the point that they treat their suppliers as part of the company and let them serve as the company warehouse. This eliminates carrying costs. For example, by using sophisticated AI-based coding techniques, Wal-mart is reportedly able to monitor items on its shelves and synchronize the rate at which it receives new stocks from suppliers with the rate at which the items are selling.

Similarly, every week the Ohio-based Bailey Controls electronically transmits to Future Electronics, based in Montreal, Canada, its latest forecast of the materials that it will need in six months so that Future Electronics can ensure that they will be available. Other well-known examples of great inventory managers supported by the marketing department and using AI or other cutting-edge computer technology are Levi Strauss & Company and Milliken & Company.

TREND ANALYSIS

Marketing researchers are always wary of committing errors by, e.g., not taking into consideration possible responses for a significant portion of their sample who did not respond (this is called nonresponse bias). As a result, they employ several techniques to reduce this error—one of these is trend analysis.

The idea is to try to discern a trend in responses between waves of respondents, e.g., early and late respondents. If a trend is found, it's projected onto nonrespondents to estimate their position on the characteristic of interest. As you probably already realize, pattern recognition techniques are excellent (and widely used) for this task. They have significantly improved the accuracy of forecasts in which they are used for trend analysis.

COMMUNICATION

Good communication in marketing is necessary from the genesis of a product or service concept to the sale of the final offering. Moreover, the introduction of speech recognition programs in the information/data collection phase of marketing research has added a versatility unknown before. Collecting information via remote devices frees personnel to perform more human-intensive chores. This ultimately leads to significant cost savings.

The ability of a marketer to effectively communicate is most serious at the launch stage. Not only do marketers need to communicate their offerings to the world beyond the company walls, they also need to do so in a way that elicits the desired response. Primarily, they have to make sure they are communicating with the appropriate target markets—those consumers who would most likely find the offering a solution to their unmet needs. This aspect of communication is always a daunting task, but armed with a program like Claritas, Inc.'s *PRIZM,* narrowing the choice of targets has become manageable.

Once the markets are identified, marketers can use multidimensional scaling (MDS) programs to design a communication strategy that effectively positions the product or service. Useful versions of MDS are *MDSCAL 5M, PREFMAP, APM* (adaptive perceptual mapping), and *MAPWISE*. Effective positioning strategies, for example, were positioning Volvo as a safe car, Mercedes as a symbol of success, or Listerine as a mouth rinse that kills germs.

Speech and pattern recognition techniques have also been used to enhance marketing communications strategy. They allow marketers to choose words, colors, and symbols that have special meaning to consumers in the target audience and are therefore more likely to catch their attention or make the message more memorable.

Globalization is taking marketers farther and farther away from the markets they are most familiar with. Many companies make this transition by hiring people from the regions they are entering. However, hiring local talents never completely eliminates the need for American marketers to either live in foreign countries where their companies have become major players or visit for extended periods. As a result, many companies train members of their top-level marketing team in the needed foreign language. Some of the equipment used in this training uses speech recognition techniques that allow it to translate certain English words vocally and phonetically into their foreign equivalent.

Finally, the communication process is not complete until there is a mechanism for feedback. In marketing, purchase by the consumer is the desired immediate response, but good marketers also want to know what consumers think after purchase or use, whether it is a complaint or a compliment.

Most marketers use content analysis techniques to analyze verbal feedback responses, but the volume of responses can test the limits of such traditional approaches. Once again, AI-based techniques like pattern recognition come to the rescue.

MAKING DECISIONS IN UNCERTAINTY

The most pervasive marketing activity, of course, is not selling but making decisions in uncertainty. The marketer has to make decisions about type of product, price, where to sell, and how to sell—in short, the marketer has to pick a strategy. Even though all these decisions are vital to the success of a marketing program, they must be made without the benefit of perfect information and without knowing whether they will work.

Long ago marketers made such decisions on the basis of experience and intuition. Success in the current environment needs a more sturdy foundation. Armed with massive amounts of data, marketers in the current competitive world make extensive use of complicated statistical and mathematical models in making all these necessary decisions.

Not too long ago, marketers relied heavily on decision scientists, using game-theory and heuristic models to make decisions with imperfect information. The accuracy of these models is greatly enhanced when they are combined with AI techniques. For example, a more reliable segmenting model for classifying non-life insurers according to their risk exposure was developed by Bert Kramer when he combined the traditional statistical model with AI techniques.[12] This is not an isolated case by any means, it definitely represents the wave of the future.

WHAT NEXT?

The exponential rate at which computer science technology is improving and the almost equally strenuous rate at which it is being integrated into marketing makes it difficult to make specific predictions. However, it is certainly safe to predict that AI will continue to be integrated into marketing, especially as they increase their capacity to handle qualitative data. We could also see within the next five years models that are run by spoken instructions. Such systems would involve marketing managers more directly in the technical aspects of model building. At that point, we would expect models to be more accurate because the input into them would more accurately reflect what marketers do now that they cannot correctly describe to the model builder or what they wish that the models would *really do*.

---

[12]Kramer, B. "N.E.W.S.: A Model for the Evaluation of Non-life Insurance Companies," *European Journal of Operations Research,* 98, 2 (1997): 419-31.

REFERENCES

Ben-Akiva, M. and Lerman, S. R. *Discrete Choice Analysis: Theory and Application to Travel Demand,* 3d. ed. Cambridge, MA: MIT Press, 1989.

Carroll, J. Douglass and Green, S. Paul. "Psychometric Methods in Marketing Research: Part I, Conjoint Analysis," *Journal of Marketing Research,* 32 (November) 1995: 385-91.

Curry, B. and Moutinho, L. "Intelligent Computer Models for Marketing Decisions," *Management Decisions,* 32, 4 (1994): 30-36.

Davies, F., Moutinho, L., and Curry, B. "Construction and Testing of a Knowledge-Based System In Retail Bank Marketing," *International Journal of Bank Marketing,* 13, 4 (1995): 4-15.

Del Valle, C. "They Know Where You Live—and How You Buy," *Business Week* (Feb. 7) 1994: 89.

Englis, B. G. and Solomon, M. "To Be and Not to Be: Lifestyle Imagery, Reference Groups, and the Clustering of America," *Journal of Advertising,* 24, 1 (1995): 13-29.

Fenn, J. and Hodgdon, P. "How Emerging Artificial Intelligence Technologies will Affect Direct Marketing," *Direct Marketing,* 58, 4 (1995): 28-31.

Fish, K. E., Barnes, J. H., and Aiken, M. W. "Artificial Neural Networks: A New Methodology for Industrial Market Segmentation," *Industrial Marketing Management,* 24, 5 (1995): 431-39.

Goodman, S. and Plouff, G. "Neural Network Modeling: Artificial Intelligence Marketing Hits the Non-Profit World," *Fund Raising Management,* 28, 4 (1997): 16-18.

Kinnear, T. C. and Taylor, J. R. *Marketing Research: An Applied Approach,* 5th ed. New York: McGraw Hill, 1996.

Kleege, S. "AT&T to Refine its Marketing Through Artificial Intelligence," *American Banker,* 158, 165 (1993): 114-15.

Kotler, P. *Marketing Management: The Millennium Edition,* Englewoood Cliffs, New Jersey: Prentice-Hall, 2000.

Kramer, B. "N.E.W.S.: a model for evaluation of non-life insurance companies," *European Journal of Operations Research,* 98, 2 (1997): 412-61.

Malhotra, N. *Marketing Research: An Applied Orientation,* 3d. ed. Englewood Cliffs, New Jersey: Prentice-Hall, 1999.

Montgomery, D., Swinnen, G., and Vanhoof, K. "Comparison of Some AI and Statistical Classification Methods for a Marketing Case," *European Journal of Operational Research,* 103, 2 (1997): 312-26.

Morrison, Ann M. "Dream Maker; the Rise and Fall of John Z. DeLorean," *Fortune* (Sept. 19, 1983): 193-97.

Moss, S. and Edmonds, B. "A Knowledge-based Model of Context-Dependent Attribute Preferences for Fast Moving Consumer Goods," *Omega,* 25, 2 (1997): 155-70.

Tucker, M. J. "Poppin' Fresh Dough," *Datamation,* 43, 5 (1997): 50-57.

Vaas, L. "Earn E-Customers' Love—E-Businesses Are Wooing Visitors with Smart Tools, But Customer Love Must be Built on Solid Data," *PC Week,* (Feb., 14, 2000): 69.

Venugopal, V. and Baets, W. "Neural Networks and Statistical Techniques in Marketing Research: A Conceptual Comparison," *Marketing Intelligence & Planning,* 12, 7 (1994): 30-39.

Wezel, Michiel C. van and Baets, W. R. J. "Predicting Market Response with a Neural Network: The Case of Fast Moving Consumer Goods," *Marketing Intelligence & Planning,* 13, 7 (1995): 23-31.

CHAPTER 8

Expert Systems and Neural Networks in Manufacturing*

As you're already aware, an expert system (ES) is a computer system composed of both hardware and software that can perform reasoning using a knowledge base. The ES uses a rule base to generate suggestions. Users input facts and preconditions that trigger the rules to provide results.

Expert systems are applied in many aspects of manufacturing because much factory work has shifted to knowledge work (such as planning, designing, and quality assurance as well as work formerly done by labor, such as machining, assembling, and handling). It's estimated, in fact, that knowledge work accounts for about two-thirds of total manufacturing cost. Robotics, ESs, and other information systems improve the productivity of labor. For example, an ES implemented at Northrop Corporation, a major producer of jet fighter plans, is responsible for planning the manufacture and assembly of up to 20,000 parts that go into an aircraft. A parts designer enters a description of the engineering drawing of a part, and the ES tells him what materials and processes are required to manufacture it. This particular system actually improves the productivity of designing parts by a factor of 12 to 18. Without the help of the ES, the

---

*Chung J. Liew, Ph.D., consultant and professor of decision sciences at the University of Central Oklahoma, contributed this chapter.

same task would require several days instead of several hours. Here are two other examples of the ES at work.

EXAMPLE ONE: EXPERT SYSTEMS INTEGRATION DRIVES VAN CONVERSIONS[1]

Mark III is the world's largest van conversion company (Ocala, FL). The company converts old vans and pickup trucks by stripping them of everything except the engine, body, chassis, seats, and air conditioning. Then they are converted into luxury vehicles with plush interiors and high-tech sound systems, usually to a customer's order. Luxuries may include Nintendo games, cocktail bars, vacuum cleaners, and fancy computers.

Every day more than 300 vehicles enter the plant. For many years, the data on each was entered by hand, creating a two to three day paperwork lag before the company even knew a vehicle had entered its parking lot. But information technology (IT) has helped change the way business is done.

Using hand-held bar-code data-collection equipment, an employee enters the vehicle identification number, location, date, and time received for each vehicle. The information is downloaded into a computer system, which integrates engineering, manufacturing, financial management, and management reporting. This system was designed to support business processes reengineering.

Using an ES, the company quickly produces a customized manufacturing order for each entering van. This allows the company to offer a large number of configurations to its dealers as well as to quickly produce special orders. For example, a German customer gave Mark III seven days to build a special van and put it on a boat bound for Germany. Not only was the company able to fulfill the order because the work started immediately, but also, because it used an integrated client/server architecture with object-oriented programming and graphical user interface, the company was able to design and follow the order moment by moment.

Using the wireless radio frequency bar codes on each van, the company tracks what is going on in each workstation automatically. The data is entered into a database that allows salespeople instant access to the status of each order. Since most of the components are manufactured in-house, it is critical for Mark III to have complete control of its processes.

---

[1]Adapted from Thomas, C. "High Tech Integration Drives Van Conversions," *APICS,* October 1993, p. 70.

Using its information systems, the company can solve any logistic, accounting, or technical problem immediately. Using a just-in-time (JIT) approach, the company assures that all parts, materials, and tools are in place when the vehicle arrives. There are no inventories because there is no need for buffer stocks.

The information system (IS) integration allows for quick and accurate payroll, which includes productivity incentive pay. It also integrates the production and inventory MRP systems with a quality-control system.

Finally, the system is used to enhance customer service after the vehicles are sold. Using the system, customer service operators can answer the more than 10,000 weekly calls from customers quickly and accurately. Having a file on each vehicle helps identify replacement parts as well. If a customer needs a replacement part, the service center finds the correct part in seconds and prepares an order for the warehouse, right from the screen. Shipment is made in less than 24 hours, compared with three days formerly.

Dealers are also better served with the new systems. For example, dealers are now reimbursed for parts and labor made under warranties in less than 10 days (it used to take 20-30 days). The results: The company is able to maintain its top-ranked position in the industry, keep profitability high, and increase production from 70,000 vehicles in 1993 (before the reengineering) to over 100,000 vehicles in 1994 (after the reengineering).

EXAMPLE TWO: GENERAL ELECTRIC'S EXPERT SYSTEM MIMICS HUMAN REPAIR EXPERTS

The Problem

General Electric's (GE) top locomotive field service engineer, David I. Smith, had been with the company for more than 40 years. He was the top expert in troubleshooting diesel electric locomotive engines. Smith was traveling throughout the country to places where locomotives were in need of repair to determine what was wrong and advise young engineers about what to do. The company was very dependent on Smith. The problem was that he was nearing retirement.

Traditional Solutions

GE's traditional approach to such a situation was to create apprenticeship teams that paired senior and junior engineers. The pairs worked together for several months or years, and by the time the older engineers finally did retire, the younger engineers had absorbed enough of their seniors' expertise to carry on troubleshooting or other tasks. This practice proved to be a good short-term solution, but GE still wanted a more effective and dependable way of disseminating expertise among its engineers and preventing valuable knowledge from retiring with people like David Smith. Furthermore, because railroad service shops are scattered all over the world, it is not economically feasible to keep an expert permanently in every location.

GE decided to build an expert system to solve the problem by modeling the way a human troubleshooter works. The system builders spent several months interviewing Smith and transferring his knowledge to a computer. The programming was prototyped over a three-year period, slowly increasing the knowledge and number of decision rules stored in the computer. The new diagnostic technology enables a novice engineer or technician to uncover a fault by spending only a few minutes at the computer terminal. The system can also explain to the user the logic of its advice, serving as a teacher. Furthermore, the system can lead users through the required repair procedures, presenting a detailed computer-aided drawing of parts and subsystems and providing specific how-to demonstrations.

The system is based on a flexible, human-like thought process, rather than rigid procedures expressed in flowcharts or decision trees.

The system, which was developed on a minicomputer but operates on microcomputers, is currently installed at every railroad repair shop served by GE, thus eliminating delays, preserving Smith's expertise, and boosting maintenance productivity.

PRODUCTION AND OPERATIONS

Production scheduling involves the allocation of raw materials and machinery to different product lines. The Goodyear Tire & Rubber Co. plant in Houston, Texas, developed an expert system, *The Manager for Interactive Modeling Interfaces* (MIMI), to improve production scheduling. Within two years, the plant had cut its inventory and operating costs, increased plant capacity, improved delivery performance, and reduced transportation costs. MIMI integrates an ES with simulation capabilities, allowing the plant to create an exact scheduling model.

An ES can also be used to build design and configuration systems that support the reuse and modification of standard designs. Two notable examples are Nippon Steel's *QDES* and Lockheed's *Clavier.*

Companies like Digital Equipment Corporation (DEC) use ESs in configuring computer systems. DEC is also using them as a salesman's assistant and to help perform scheduling. Aerospace firms like Boeing and Lockheed use ESs for navigation control, planning and scheduling, fault diagnosis, and training. Boeing even has a workstation called *Aquinas* to help acquire knowledge and build knowledge bases.

Compaq Computer's *SMART* system is an integrated call-tracking and problem resolution system that contains hundreds of diagnoses of problems that have arisen in the use of Compaq products. The Customer Service Department uses it in responding to calls to the central toll-free number. Incoming customer problems are presented to SMART, which retrieves the most similar cases from its knowledge base and presents them to the customer service analyst, who uses them to resolve the problem. When the first version of SMART was installed, the percentage of customer problems resolved on the first call rose from 50 percent without the system to 87 percent. According to Compaq, the system paid for itself within a year. SMART runs on PCs running the Microsoft Windows operating environment.

Oil companies like Schlumberger-Doll and Amoco use ESs in mineral exploration and to identify faults in the refining process. Airline companies use them to assign planes to gates at major airports. Telecommunications firms use them every day to identify cable and switch maintenance problems and to provide on-line assistance to network analysts and operators. Trucking companies use them for resource allocation and scheduling.

Financial institutions and investment firms use expert systems in providing financial and estate planning advice. Underwriters at insurance companies use them when they are deciding whether to accept applications for coverage.

Expert systems can be used to detect and analyze patterns of variation in manufacturing quality control charts. The ES looks for six potential patterns of variation (1) trend, (2) cycle, (3) mixture, (4) shift, (5) stratification, and (6) systematic. Statistical significance tests are used as interpretive rules to determine which pattern is present. Once the pattern is identified, the ES supplies the user with possible causes for the out-of-control condition, as well as the magnitude of the problem and where it starts and stops.

Many operations activities like facilities layout design, product planning and design, process selection, and CAD/CAM are also being made in conjunction with ES technology. Most successful manufacturers improve their designs and processes continually. The Assembly Automation Division of Giddings & Lewis decided to have its process planners build on the company's knowledge base instead of creating new plans for each order. Management saw cultivating the knowledge base as an expedient way to reduce time to market and respond to customer demands for installing automated custom assembly lines, such as the body line at the Saturn plant and the chassis line for the new Ford Explorer. These were done in six months instead of the usual nine to 18 months.

BUSINESS PROCESS REENGINEERING

A good example of ES applications in business process reengineering is the *Prism* telex classification system developed by Cognitive Systems, Inc. Prism is used in banks to route incoming international telex communications to appropriate recipients, a task relying on accurate classification of the telex. Prism has been found to increase classification accuracy from 75 percent to 90 percent and reduce average time for routing a telex from several minutes (with human telex operators) to 30 seconds. Prism also demonstrated improvements in accuracy and speed over a previous rule-based implementation of the same type of system. Prism runs on Macintosh-11 workstations.

Another example of business process reengineering is NEC Corporation's *SQUAD* system. SQUAD is an ambitious long-term effort to provide a corporate-wide system for capturing and distributing software quality control knowledge. Some 3,000 cases per year have been added to the system since 1982. The developers estimate the productivity savings due to the use of SQUAD at over $100 million per year. SQUAD runs on UNIX workstations.

ROBOTICS

A robot is a device that mimics human actions and appears to function with some degree of intelligence. A robot with AI capability is an electromechanical manipulator able to respond when it perceives a change in its environment. The sensory subsystem is programmed to "see" or "feel" the environment. For example, an industrial robot can manufacture one of many parts in its repertoire and manipulate it to inspect it for defects, recognizing very small departures from established standards. Robots have been used extensively in Japan to improve the quality and reduce the cost of their products. They are reliable, consistent, accurate, and insensitive to hazardous environments.

Robots are commonly used where it would be unsafe or unhealthy for a human to perform the same task. The major problem of controlling the physical actions of a mobile robot might not seem at first look to require much intelligence. Even small children are able to navigate through their environment and to manipulate items such as playing with toys, using spoons, and turning on a TV. But these tasks, performed almost unconsciously by humans, when performed by a machine require many of the same abilities used in solving more intellectually demanding problems.

Research on robots or robotics has been the source of many AI applications. It has led to techniques for modeling "the state of the world" and for describing the process of change from one world state to another. It has led to a better understanding of how to generate action plans in sequence and how to monitor their execution. One continuing challenge is to develop methods for planning at high levels of abstraction, ignoring details, and then for planning at lower levels when details become important.

NEURAL NETWORKS IN MANUFACTURING

The human is the most complex computing device known. The brain's powerful thinking, reasoning, creation, remembering, and problem-solving capabilities have inspired many scientists to attempt computer modeling of its operation. Some researchers have sought to create a computer model that matches the functionality of the brain in a very fundamental manner—the result has been neural computing.

The neuron is the fundamental cellular unit of the nervous system and the brain. Each neuron functions as a simple microprocessor that receives and combines signals from many other neurons through input processes called dendrites. If the combined signal is strong enough, it activates the firing of the neuron, which produces an output signal. The output signal moves along a component of a cell called the axon. This simple transfer of information is chemical but has electrical side effects that we can measure.

The brain consists of hundreds of billions of neurons loosely interconnected. The axon (output path) of a neuron splits up and connects to dendrites (input path) of other neurons through junctions called synapses. Transmission across a synapse is chemical and the amount of signal transferred depends on the amount of the chemical (the neurotransmitter) released by the axon and received by the dendrites. This synaptic efficiency is what is modified when the brain learns. The synapse combined with the processing of information in the neuron form the basic memory mechanism of the brain.

In an artificial neural network, the unit analogous to the biological neuron is referred to as a "processing element." It has many input paths whose values it combines, usually by a simple summation. The result is an internal activity level for the processing element. The combined input is then modified by a transfer function, which can be a threshold function. This threshold function may only pass information if the combined activity reaches a certain level, or it may be a continuous function of the combined input.

The output path of a processing element can be connected to input paths of other processing elements through connection weights that correspond to the synaptic strength of neural connections. Since each connection has a corresponding weight, the signals on the input lines to a processing element are modified by these weights prior to being summed. The result is a weighted summation.

Thus, a neural network consists of many processing elements joined together as described. Processing elements are usually organized into groups called layers or slabs. A typical network consists of a sequence of layers or slabs with full or random connections between successive layers. There are typically two layers with connections to the outside world (1) an input buffer where data is presented to the network and (2) an output buffer which holds the response of the network to a given input. Layers distinct from the input and output buffers are called hidden layers. Applications of NNs are language processing (text and speech), image processing, character recognition (hand writing recognition and pattern recognition), and financial and economic modeling.

APPENDIX I

Artificial Intelligence Software

Following is a list of popular AI, ES, and NN software products. For pricing information and an updated product description, contact the vendor.

AI Ware
3659 Green Road
Beachwood, Ohio 44122
Tel: 216-514-9700
Fax: 216-514-9030
e-mail: *ai-sales@aiware.com*

Business Advisor
A complete forecasting and optimization decision support system providing solutions to business problems.

CAD/CHEM
A formulation modeling and multi-objective optimization system with intuitive neural networks, sensitivity analysis, data clustering, and fuzzy objective functions.

Process Advisor
A process monitoring and optimization system.

Acquired Intelligence Inc.
1095 McKenzie Avenue, Suite 205
Victoria, British Columbia
V8P 2L5 Canada
Tel: 250-479-8646
Fax: 250-479-0764

Acquire
An expert-based knowledge acquisition system with business applications.

BioCamp Systems, Inc.
4018 148th Avenue, N.E.
Redmond, Washington 98052
Tel: 800-716-6770
Fax: 425-869-6850
Web site: *http://www.bio-comp.com*

Neuro Genetic and Trade
A NN program to discover and model hidden important relationships in data such as sales figures, financial market information, marketing research survey information, customer profiles, and demographic information.

California Scientific Software
10024 Newton Road
Nevada City, California 95959
Tel: 800-478-8112
Fax: 916-478-9041
e-mail: *sales@calsci.com*
Web site: *http://www.calsci.com*

Brain Maker Professional
NN simulation software used to solve business problems; forecast revenue and costs; conducts manufacturing analysis; evaluate processing costs; analyze loan applications; analyze investments; predict current prices, bond ratings, and the S&P 500 index; appraise real estate; perform market analysis; and generate financial indicators.

Elf Software Company
210 W. 101st Street
New York, New York 10025
Tel: 212-316-9078

Access Elf
A query interface software to access Microsoft databases.

HNC Software Inc.
5930 Cornerstone Court West
San Diego, California 92121
Tel: 619-546-8877
Fax: 619-452-6524

Falcon Payment Card Fraud Detection System
NN software that audits and detects fraudulent transactions by customer account of card-issuing financial institutions.

Capstone Application Decision Processing System
Software that books new payment card accounts.

IBM
Department AC 297, AS/400
P.O. Box 16848
Atlanta, Georgia 30321
Tel: 800-IBM-CALL
Fax: 800-2 IBM-FAX

Knowledge Tool for AS/400
A troubleshooting expert-based application software performing risk analysis and capacity planning.

Integrated Reasoning Shell for OS/2 Release 3
An ES shell providing for the development of knowledge-based applications in a workstation environment.

Neural Network Utility Product Family
NN software used to identify financial trends and patterns to guide business operations.

Level 5 Research
1335 Gateway Drive, Ste. 2005
Melbourne, Florida 32901
Tel: 800-444-4303
Tel: 407-729-6004
Fax: 407-727-7615

Level 5 Object Professional Release 3.0 for Microsoft Windows
A knowledge-based ES shell tool.

Level 5 Quest
A fuzzy logic base-query environment search engine.

Multi Logic
1720 Louisiana Blvd., N.E., Suite 312
Albuquerque, New Mexico 87110
Tel: 800-676-8356
Fax: 800-256-8356

Multi Logic Exsys Professional
A knowledge-based NN development tool used for financial modeling and allocation of corporate resources to improve rates of return.

Neural Systems, Inc.
2827 West 43rd Avenue
Vancouver, British Columbia
V6N3HG Canada
Tel: 604-263-3667

Genesis
A neural network for business decision-making involving a multitude of variables.

Neural Ware Inc.
202 Park West Drive
Pittsburgh, Pennsylvania 15275
Tel: 800-635-2442
Fax: 412-787-8220
Web site: *http://www.neuralware.com*

Neural Works Predict
An NN used to solve prediction, modeling, and classification problems.

Scientific Consultant Services, Inc.
20 Stagecoach Road
Selden, New York 11784
Tel: 516-696-3333

N-Train: Neural Network System
NN software for solving business problems.

The Trading Simulator
A NN that simulates trading accounts and portfolios.

Sterling Wentworth Corp.
57 West 200 S., Suite 500
Salt Lake City, Utah 84101
Tel: 800-752-6637

Expert Series
An ES used in personal financial planning with clients.

Teranet
1615 Bowen Road
Manaio, British Columbia
V9F 1G5 Canada
Tel: 604-754-4223

ModelWare Professional
A modeling software pogram used by a company considering the various business factors it faces.

Texas Instruments Corp.
P.O. Box 660246
Mail Station 8671
Dallas, Texas 75166
Tel: 800-336-5236

Capital Investment Expert System
AI software that analyzes, manages, and reports on options for purchase of machinery and equipment.

Triant Technologies
20 Townsite Road, 2nd Floor
Nanainmo, British Columbia
V9S 5T7 Canada
Tel: 800-663-8611
Fax: 604-754-2388
e-mail: *mail@triant.com*

Model Ware
An ES used for predictive modeling.

Glossary

ADAPTIVE
Ability of software to perform well in a changing business environment.

ADAPTIVE FUZZY ASSOCIATIVE MEMORY (AFAM)
Fuzzy reasoning system that adjusts to the central measure of the fuzzy space as processing continues. Both the fuzzy set topology and the rules composition are modified. One type of AFAM generates the rule set from approximated fuzzy sets and actual data streams.

ADAPTIVE FUZZY SYSTEMS (AFS)
Systems working at two levels. At the local (lower) level, a neural network approximates patches or rules. At the global (higher) level, the patches approximate the entire system. AFS learn from experience and program themselves; no human expert enumerates the rules—data flows into a neural or statistical system to generate them. AFS serves as the human expert. It learns from experience with new data refining the knowledge base. One application of AFS is in production and delivery movement of products and delivery trucks.

Adaptive fuzzy systems are a brainlike neural network, a computer system imitating how brains learn and recognize patterns and formulating fuzzy rules from training data. Experience is the basis for the learning process. Data is inputted, rules are outputted. Pattern recognition is the basis for the rules being generated. Fuzzy sets are the patterns, and fuzzy rules are the relations. The fuzzy system uses the rules to reason with the patterns. The more data used, the better the rules derived. An AFS changes or fine tunes its rules as it samples new data. Initially, the rules change quickly. However, with more samples, the rules change more slowly.

ALPHA-LEVEL
Measure of the minimum degree in a fuzzy set or fuzzy region. A truth membership value is the alpha-threshold level. Strong and weak are the two kinds of alpha thresholds.

AMBIGUITY
Vagueness or lack of clarity. An ambiguous reference involves a control variable, statistic, or other symbol whose meaning has several distinct possible interpretations. Ambiguity is at the heart of fuzzy ontology. Ambiguity relates to the notion of undecidability in fuzzy regions.

ANTECEDENT
Conditions coming before the present model state. In a rule-based expert system antecedent refers to the premise for a rule. In a fuzzy system, it refers to the combined truth state of the current fuzzy region for a specific control or solution variable.

APPROXIMATE REASONING
A reasoning process incorporating fuzzy sets, fuzzy operators, hedges, and decomposition rules in an implication process. The final membership function applicable to a fuzzy consequent space is formulated by processing each premise fuzzy assertion with its combining operators and applying the suitable composition rule. The combining operators are AND and OR. With the decomposition method, the fuzzy space is used to formulate the solution variable referred to as the expected value. *Defuzzification* is the term associated with this process.

APPROXIMATION
Process of creating a fuzzy set from a scalar or nonfuzzy continuous variable. With a scalar variable, there will typically be a bell-shaped fuzzy space. Approximations can usually be used for scalars, fuzzy

sets, and fuzzy regions. Hedges around, close to, or about can approximate the fuzzy set or scalar.

ARTIFICIAL INTELLIGENCE (AI)
Application of human reasoning techniques to machines. AI systems use sophisticated computer hardware and software to simulate the functions of the human mind, including reasoning, sensation, perception, learning, and communicating, in order to think and solve complex business problems. Computers are made to act intelligently. AI applications include forecasting stock market prices via sequence prediction planning. Plans describe many alternative sequences of actions, specifying conditions based on the sequences that were followed. Program-like structures such as fuzzy algorithms are used to represent plans. Pattern recognition applications also exist. Further, applications are aided by long chains of if-then rules.

ARTIFICIAL LANGUAGES
Mathematically defined classes of signals that can be used for communicating with machines. Usually referred to as programming languages.

ASSERTIONS
Fuzzy propositions. Assertions may be conditional or unconditional. Conditional fuzzy assertions update the solution space with a strength (degree) that depends on the truth of their predicate, or conditional part. An example of a conditional state is "if our price is below our competitor's, our sales volume will increase." Unconditional assertions set up a fuzzy region that serves as support for the solution space. An unconditional fuzzy statement is "our price should not exceed the prevailing market price." A fuzzy system model is comprised almost fully of assertions (or fuzzy propositions).

ASSIGNMENT
Allocating values to variables subject to constraints.

ASSOCIATIVE MEMORY
A system storing data in parallel form. Data is searched based on some characteristic or feature. Patterns of information are noted.

AUTOMATIC PROGRAMMING
Software that automatically writes other software to be used in artificial intelligence. This program can take in a very high-level description of what the program is to achieve and generate a program in a particular programming language. Debugging may be used as a problem-solving strategy.

BACKWARD CHAINING
Backward reasoning by AI systems using goals, rules, and facts.

BAYES THEOREM
A probability theorem specifying when the a priori probability of a hypothesis is known before evaluating the test results. The future test results can be used to adjust the cumulative probability to predict the a posteriori probability. In practice, this is a measure of subjective probability based on a rational belief of the probability of an event taking place. The scale range is from total disbelief (zero) to total belief (one).

BELIEF SYSTEM
A system establishing a reasonableness state for an object or a specified model state. This involves projecting its domain into a set of candidate values. A belief system relates to the subjective intentions of the observer in confirming or denying the validity of a model assertion or proposition.

BIT VALUE
Zero or one.

BIVALENCE
The opposite of fuzziness, allowing there are two ways to answer each question, True or False, one or zero.

BIVALENT LOGIC
Each statement is true or false or has associated with it the truth value of one or zero.

BOT
Short for robot.

Glossary

BOTTOM-UP
In AI, human experts with very simple rules. Complex behaviors like intelligence result from a multitude of many parallel applications and interaction of these simple rules.

BUSINESS PROCESSING MODELING (BPM)
Functional decomposition of an expert's activities into a set of tasks, policies, or functional units.

CARTESIAN PRODUCT (CP)
The product of n-state spaces and m-state spaces resulting, in a state space of (m x n) values. CP represents a set of all the combinations of two or more value sets.

CASE-BASED EXPERT SYSTEM
A system using an inductive knowledge method to conduct expert system reasoning. The index library is used to efficiently search and retrieve the cases most similar and relevant to the current problem. An adaptation module creates a solution for the current problem by modifying a prior solution (structural adaptation) or creating a new solution based on similar processes used in prior cases (derivational adaptation).

A case-based expert system contains many cases which have different results. Cases consist of information about a situation, the solution, results of using that solution, and key attributes. The expert inference engine will search through the case base to find the historical case that best matches the characteristics of the problem to be solved. The solution of a matched historical case will be modified and used as a suggestion for the current problem.

CASE-BASED REASONING
Drawing suggestions or conclusions based on previous scenarios and resulting actions, from which future moves can be deduced.

CENTROID METHOD
A defuzzification approach that results in the expected value for a consequent variable by computing the gravity center associated with the consequent fuzzy region.

CHAOS
A random state but with deterministic behavior. The property of chaos is that if you select any two beginning points of a chaotic system, regardless of how close they are they will result in two paths that will diverge over time.

CHROMOSOME
Weights in a network gathered together into an array. The chromosome can be appraised for any training example: An error for that example is determined. The error equals the absolute value or square of the difference between the computed value and the actual value. The fitness function is the sum of the errors for all the examples in the training set. The genetic algorithm can minimize the fitness function using selection, crossover, or mutation.

CLUSTERS
Data groupings indicating patterns.

COGNITIVE STRUCTURE
An agent's model coupled with its mechanism for using that model in selecting actions. The cognitive structure often includes an agent's goals, intentions, and executable actions.

COMPATIBILITY
The central idea of fuzzy set theory associated with the degree to which a control variable's value is compatible with the related fuzzy set. Some believe that compatibility is the essence of fuzzy logic and possibility theory. The extent of membership value returned by appraising the assertion is a measure of this compatibility.

COMPATIBILITY INDEX (CI)
A real number between zero and one in a fuzzy model showing the degree to which the control variables are compatible with the underlying fuzzy sets. The index is the truth membership function that applies to the consequent fuzzy space. The CI gives a measure of fuzzy compatibility when the index ranges between 0.2 and 0.8. Numbers higher or lower than this reveal whether the control variables are at the extremes of the combined fuzzy sets.

CI measures the compatibility of the fuzzy model to the model data. It looks at the factors associated with a model, including the strength and consistency of its recommendations and their relationship to the model's logic—How are the rules responding to the model data? We look at the structure of the fuzzy sets to each solution variable to find out if the model is working appropriately.

The two types of model compatibility are statistical and unit. Statistical compatibility measures the model's performance over a wide data range. Unit compatibility measures the recommendation strength of a single execution of the model. One unit compatibility index exists for each solution variable in a model or policy. Should a low unit compatibility affect confidence in the model results? In general, the more rules contributing to a solution, the less pronounced the compatibility problems. The unit CI is a key consideration when deciding whether to accept a model's recommendation or whether to modify the recommendation based on the strength of the production rule set.

COMPENSATORY OPERATORS

These make it possible to combine fuzzy propositions or assertions with AND and OR operators.

COMPLEMENT

The negation of the fuzzy set. The fuzzy complement reveals the extent to which an element is not a member of the fuzzy set. The complement also has members that have partial exclusions.

COMPOSITE MASS

A defuzzification approach generating the expected value for a consequent variable by looking at the area of the fuzzy consequent with the highest intersection density of premise fuzzy sets. This is the area where most of the rules are executed so as to provide the most votes for a value from this region. The composite mass decomposition technique applies the evidence rules in deriving a value for the solution variable.

COMPOSITE MAXIMUM (CM)

A defuzzification approach resulting in the expected value for a consequent variable by appraising the edges of the fuzzy space across the fuzzy region's domain. CM selects the point having the maximum truth value along the edge and uses the domain value at that point as the solution variable. When there is a double-edged plateau for the region, the plateau's center is chosen.

CONJUNCTION
Applying the intersection operator (AND) to two fuzzy sets. For every point on the fuzzy set, use the minimum of the membership functions for the two intersected sets.

CONJUNCTIVE NORMAL FORM
A formula written as a conjunction of clauses.

CONNECTIONISM
Review of computer programs inspired by neural systems.

CONSEQUENT
The action aspect of a rule. It defines each control fuzzy region that updates a particular solution variable.

CONSISTENCY PRINCIPLE
The principle of probability theory stating that the likelihood (possibility) of an occurrence is at least as great as its probability.

CONSULTING SYSTEMS
Systems that try to answer users' queries by asking questions of the user or of a database about the truth of propositions they may know about.

CONTAINMENT
The degree to which one thing (e.g., a set) contains another thing (e.g., a set).

CONTINUOUS VARIABLE
A series of nondiscrete possible values used as a control variable.

CONTRAST
A hedge operation moving the truth function of a fuzzy set around the 0.5 degree of membership.

CORRELATION ENCODING
Rules coupling the truth of the consequent assignment fuzzy region with the truth of the premise of the rules, based on the notion that the consequent truth cannot exceed the premise truth.

CORRELATION MINIMUM RULE
An implication method correlating the consequent fuzzy region with the antecedent (premise) fuzzy truth. The membership value of the

current consequent fuzzy region is the minimum of the fuzzy region and the truth of the premise.

CORRELATION PRODUCT RULE
An implication method correlating the consequent fuzzy region with the antecedent (premise) fuzzy truth by taking the outer product of the membership values.

CRISP SET
A set in which membership is either totally within or totally excluded. A property of this set is well-defined membership behavior.

DATABASE MANAGEMENT SYSTEM
In expert systems, controls input and management of both the knowledge and domain databases.

DATAFLOW
Architecture for parallel processing systems in which instructions execute whenever data flows to them, with data usually flowing to multiple instructions at once.

DECISION SUPPORT SYSTEM (DSS)
Computer-based software that helps decision makers by providing data and models. It conducts primarily semistructured tasks. A DSS does not make decisions but merely attempts to improve decisions by providing indirect support without automating the entire decision process.

DECOMPOSITION METHOD
An approach used to formulate the expected value of a model solution variable from a consequent fuzzy region. The method selected varies depending on the type of expectation anticipated in the composite fuzzy region.

DEDUCTIVE RULE-LEARNING
A system that deduces additional rules from known domain rules and facts.

DEFUZZIFICATION
An approach to formulate a scaler representing a control variable's expected value from a fuzzy set.

DELPHI METHOD
A consensus-gathering approach used to appraise a premise and its possible outcomes, used in gathering and validating knowledge when experts provide information for an expert system or fuzzy model.

DEMPSTER-SHAFER THEORY
A theory responding to a hypothesis with mutually exclusive possible solutions. The evidence is analyzed.

DIFFERENCE OPERATORS
Those parts of software that identify types of variations between objects.

DIGITAL SIGNAL PROCESSING
The aspect of information science that converts analog measurements into lists of digits or numbers and then uses math to transform these lists into new lists of numbers.

DILUTION
A hedge activity, typically expressed as *somewhat, quite,* or *rather,* moving the truth function of the fuzzy set so membership values are increased for each point along the line. As a result, for a particular variable's membership rank, its value in a diluted fuzzy set will increase.

DISJOINT FUZZY SPACE
Consequent fuzzy space having a discontinuous region. There may be prohibited zones requiring special defuzzification.

DISJUNCTION
Application of the union operator to two fuzzy sets. For every fuzzy set point, the maximum of the membership functions is used for the two combined sets.

DISTRIBUTED ARTIFICIAL INTELLIGENCE
A system in which several agents coordinate their activities to achieve common goals.

DOMAIN
The range of real numbers applying to the mapping of a fuzzy set. It can be any set of negative or positive monotonic numbers.

DOMAIN DATABASE
The set of facts and information relevant to an area of interest (the domain).

DYNAMIC DECISION NETWORK
A belief network used to select utility-maximizing actions. It uses probabilistic inference.

DYNAMICAL SYSTEM
A system changing over time. The rate of change may be a function of time or of system constraints.

EDGE
Any boundary between parts of the image with markedly different values of some property, such as intensity.

ELEMENTHOOD
A property to some degree contained in the elements. Elementhood puts itmes in boxes. Elementhood holds within sets.

EMERGENT BEHAVIOR
Interaction between a machine and the environment.

ENTROPY
Uncertainty or disorder in a system.

EVOLUTIONARY SYSTEM
A machine, program, or procedure that creates submachines based on their ability to conduct tasks, such as solving problems or recognizing patterns in a real world situation.

EXPECTED VALUE
When defuzzifying a fuzzy region, the value representing the model's value for a solution or control variable. This value constitutes the central measure of the fuzzy space based on the defuzzification method used.

EXPERT SYSTEM (ES)
A program incorporating reasoning, inference, logical, or if-then functions to solve a business problem, such as how to reduce a specific cost or improve productivity. The ES provides not only a rec-

ommendation but also the logic it used to reach its decision. Thus, the ES interacts with users in formulating optimal decisions. Like a human expert, an ES provides advice by drawing on its own store of knowledge and by requesting information unique to the problem at hand. After the system has adequate information, an answer or result is returned to the user.

ESs are more suitable for unstructured tasks. A problem is suitable for an ES if it requires the use of expert knowledge, judgment, and experience. ESs can be used in transaction processing, determining the adequacy of expense provisions and revenue sources, scheduling, routing, financial analysis, competitive analysis, report preparation and analysis, risk evaluation, appraisal of internal control, credit authorization, claim authorization and processing, strategic planning, strategic marketing, manufacturing and capacity planning, and resource planning.

Expert systems need huge databases of information. Two components of ESs are the knowledge base and the inference engine. The knowledge base is the trees of bivalent rules. The inference engine is a method of reasoning or "chaining" with the rules. A forward chain predicts effects when it has been given the causes. It answers the what-if question. A backward chain searches for causes based on known effects. It answers the why question.

Fuzzy systems are a kind of ES because they also store knowledge as rules in the form of fuzzy rules or patches. ES work with logic or symbols; fuzzy systems work with fuzzy sets and have a numerical or math basis facilitating math analysis and simple chip design.

Figure G-1 presents the basic structure of an expert system.

EXPERT SYSTEM SHELL

A program that contains all the important elements of an expert system except domain-specific knowledge. For an ES shell to be used successful, the domain characteristics must match those the shell's internal model expects. An ES shell is basically a collection of software packages and tools used to design, develop, implement, and maintain expert systems.

Glossary

Figure G-1. Basic Structure of an Expert System

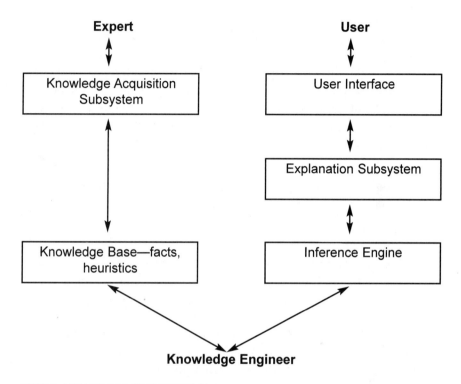

EXPLANATION SUBSYSTEM

A unit of expert systems that evaluates the structure of the reasoning conducted by the system and explains it to the user.

FAT THEOREM

A fuzzy approximation theorem; the premise that any system can be replaced with a fuzzy one. In mathematical terms, the theorem indicates that a fuzzy system with a finite rule set can uniformly approximate any continuous system. The system has a graph or curve in the space of all input and output combinations. Each fuzzy rule defines a patch in this space. A patch widens when the uncertainty increases.

FIT VALUE
A degree or number between zero and one.

FIXED POINT ATTRACTOR
The limit point associated with a dynamic system. Neural networks encode patterns as the fixed points on neural dynamic systems. Inputs or questions then surround the closest stored pattern. Some neural nets put computational solutions at fixed points in order to find a low-cost combination of resources or a low-cost schedule or route. A fuzzy system can use a neural network to find fuzzy rules by typing data clusters to neural fixed points.

FORWARD CHAINING
Reasoning that proceeds forward from facts, chains through rules, and finally sets forth the goal.

FUZZINESS
Multivalence in which there are three or more, perhaps even an infinite spectrum of, options, rather than just two extremes. Fuzziness means analog rather than binary. Fuzziness is the gray areas between black and white; it relates to the degree of imprecision or vagueness associated with a process or property in daily events or happenings. One application of fuzziness is improving production management by making machinery smarter by reasoning to improve efficiency. Machine intelligence results in a better product.

FUZZY ASSOCIATIVE MEMORY (FAM)
An array of fuzzy conditions or actions. FAM describes how a fuzzy system works.

FUZZY COGNITIVE MAP (FCM)
A fuzzy reflection of the world in the form of a causal picture. There is an edge or arrow that is a fuzzy rule between fuzzy sets. It can learn causal patterns. An expected outcome is estimated on the basis of the input. It ties facts, things, and processes to values, policies, and objectives to predict how complex situations or occurrences will interact and develop. FCMs are significantly aided by feedback. One possible application is plant control.

FUZZY ENTROPY

Measurement of the degree of fuzziness of a fuzzy set. The closer to the midpoint, the fuzzier the set. The closer to a corner, the less fuzzy. The measure of fuzziness ranges from 0 percent for a nonfuzzy set to 100 percent for a purely fuzzy set. The more a fuzzy set or description resembles its own opposite, the fuzzier it is.

FUZZY KNOWLEDGE ENGINEERING

Extraction of decisions based on information obtained from subject knowledge experts, articles, procedure manuals, and reference documents.

FUZZY (CLOUDY, VAGUE) LOGIC

A type of mathematics involving reasoning with fuzzy sets. It deals with nonprecise values with a certain degree of uncertainty. Fuzzy logic furnishes the basis for designing and writing fuzzy models. It is an approach to encoding and using imprecise information. As multi-valued or "vague" logic it infers that everything is a matter of degree such as truth. The approach simulates the process of normal human reasoning by allowing the computer to behave less precisely and logically than conventional computers. Fuzzy logic provides flexibility, options, imagination as well as allowing for observation.

FUZZY MODEL

The synthesis of expert knowledge in the form of rules and fuzzy sets with their associated machine representations. It requires an understanding of how the model is expressed in procedural code, how the components are encoded, and the formulation of the rules, model parameters, and fuzzy sets. Each rule in a fuzzy model specifies a relationship between a predicate fuzzy region and a fuzzy region applicable to one of the solution variables. When a rule is evaluated the solution variable is updated. The results of a fuzzy model represent the combined action of many rules executed in parallel form.

FUZZY NUMBERS

Numbers with fuzzy properties. Fuzzy numbers follow conventional arithmetic and also have the ability to subsume each other.

FUZZY OPERATORS

Connecting operators such as AND and OR that combine antecedent fuzzy propositions to result in a composite truth value. Fuzzy operators may use the minimum-maximum rules. Fuzzy operators set forth the nature of fuzzy logic for an implementation.

FUZZY PRINCIPLE
The principle that everything is a matter of degree.

FUZZY REASONING
Software and chips designed to make computers reason more like people. This makes them smarter and easier to work with.

FUZZY RULES
A condition taking the form of if . . . then, such as "if A is X, then B is Y." X and Y are fuzzy sets. A rule is the relationship between fuzzy sets. Each rule defines a fuzzy patch (equal to X multiplied by Y). The greater the distance between X and Y, the wider and more uncertain is the fuzzy patch. If there is more certain knowledge, there will be smaller patches or more precise rules. Fuzzy rules represent the knowledge building blocks in a fuzzy system. Each fuzzy rule serves as an associative memory relating a fuzzy response Y to a fuzzy stimulus X. Fuzzy rules can be developed using neural networks. They learn from experience based on prior data.

A depiction of a possible process to derive fuzzy rules follows:

Data ⟶ Neural Network ⟶ Fuzzy Rules

FUZZY SETS
Building blocks of all fuzzy systems. Fuzzy sets arise when a set partly contains an element, such as when department personnel include highly motivated employees or employees who are satisfied with their positions. A nonfuzzy set is when members are all or none. A set of even numbers has no fuzzy members.

FUZZY SPACE
The region in a model consisting of intrinsic fuzzy properties. Fuzzy spaces relate to the fuzzy sets created by set theory operations and the consequent sets generated by the approximate reasoning mechanism.

FUZZY SYSTEM
A system storing hundreds or thousands of common-sense rules provided by an expert. Each new piece of information activates all the fuzzy rules to some extent. There is a set of fuzzy rules converting

inputs to outputs. Then the fuzzy system refines the outputs to result in a final output or answer. Fuzzy systems are smart. A typical application is one controlling production runs. A fuzzy machine can have sensors to measure the production load. The system acts at a higher cognitive level to reason with the fuzzy rules. Fuzzy systems can model dynamic systems that change over time. Fuzzy systems can be dumb or smart depending on the fuzzy rules, and the brain's software or training and experience. In an easy case, an expert specifies the rules in words or symbols. In a sophisticated case, a neural system learns from the data rules or behavior of human experts.

GAME OF STRATEGY
In game theory, a sequence of moves each of which requires a choice between alternatives, to be made by a partial participant or the team of business professionals.

GENERALIZED DELTA RULE
An adjustment of weights.

GENERATION PROCESS
A method of generating finite descriptions or data structures for the nodes of the state space and for their interconnections.

GENERATOR
A plausibility-ordering technique that automatically derives first the most desirable option or alternative below a given situation and then less desirable alternatives.

GENETIC ALGORITHMS (GAs)
Search algorithms based on the mechanics of natural selection and genetics. They use a structured but randomized information exchange to form a search algorithm. They look at historical information to speculate on new search points with expected improved performance. In every generation, a new set of artificial creatures (strings) is created using pieces of the fittest of the old with an occasional new part tried, to enhance results. The performance of a GA depends on such factors as parameter settings (e.g., population size, crossover rate), success criteria, the operators, and method to encode candidate solutions. Business applications include machine learning, including clas-

sification and prediction; financial modeling; job-shop scheduling; price-bidding strategies; and sensors for robots. GAs work by discovering, concentrating, emphasizing, and recombining good building blocks of solutions in a highly parallel form. In effect, good solutions are made up of good building blocks.

An emerging area is using GAs to formulate aspects of neural networks, such as using network architecture in which the number of units and their interconnections can change the learning rules and weights used in a network. GAs may be used with some success in situations where there is a large search space, the space is not well understood, there is a lot of noise in the fitness function, and global optimization is not required. (Noises in the fitness function means it must take error-prone measurements from a real world process.)

GENETIC PROGRAMMING (GP)
Assurance that all expressions and subexpressions used in a program have values for all possible arguments (unless execution of an expression terminates the program). GP begins with a population of random programs, using functions, constants, and sensory inputs that are believed to be those that will be required if programs are to be effective in the domain of interest. A program is appraised by running it to determine how well it performs.

HEDGE
The transformation of one fuzzy set into another new fuzzy set by modifying the surface characteristics associated with a fuzzy set. Hedges can approximate a scalar or another fuzzy set (e.g., close to, near, about), intensify a fuzzy set (e.g, extremely, very), intensify or diffuse by contrasting (e.g., positively, typically), complement a fuzzy set (e.g., not), and dilute a fuzzy set (e.g., rather, somewhat). How hedges are defined is a crucial consideration in building and validating fuzzy systems. In a functional fuzzy system, users define their own hedge terminology and how the hedge will alter the surface of a fuzzy set.

HEURISTIC SEARCH
The way computers solve large state-space problems.

IMPLICATION
The method used to assign truth to fuzzy assertions and propositions associated with the logical implication process. Fuzzy implication operators are OR and AND, and there are rules about truth that they relate to. Most fuzzy systems have two implication rules, the bounded arithmetic sum and minimum-maximum.

IMPRECISION
A characteristic that is a measure of its overall system fuzziness in a process, concept, or event, A fuzzy system is by definition imprecise; with greater detail, imprecision diminishes.

INDUCTIVE RULE-LEARNING
A method of formulating new rules for a domain not derivable from any prior rules.

INFERENCE ENGINE
The mechanism that processes data the user inputs to find matches with the knowledge base, where the expert's information is stored. The user interface enables the user to communicate with the program. The explanation facility shows the user how each decision was derived. It contains the inference strategies and controls used by experts to manipulate knowledge and domain databases. It is the brain of the expert system.

INNOVATIVE
The ability of software to create something new and original.

INTELLIGENT AGENTS
Software that can conduct intelligent functions. The key attributes of intelligent agents are autonomy, communication ability, capability for cooperation, reasoning abilities, and adaptive behavior.

INTENSIFICATION
Hedge activity, typically pronounced, that moves the truth function of a fuzzy set so that degrees of membership values are decreased for each point along the line. Thus, the value of a variable's membership rank in the intensified fuzzy set will decrease.

INTERSECTION
The conjunction of two fuzzy sets.

INTRINSIC FUZZINESS
The property of fuzziness representing an inseparable characteristic of a process or event. The population is comprised primarily of fuzzy processes representing measurable events that cannot be separated into distinct groupings.

INVENTION
The description of an object to perform a described task. The task to be achieved may relate to an operation that maps situations. The task description may relate to an explanation of the function values on specified inputs. The invention generated from problem-solving can correspond to a finite description of a function carrying out the task. However, an invention should not be designed to depend on another.

KALMAN FILTER
The best math algorithm to forecast the current state of a linear system based on a sensor measurement. An optimal predictor provides the best estimate of the next or future state of the system given all current and previous system measurements.

KNOWLEDGE ACQUISITION FACILITY
A method of interactive processing between the system and the users; how the system acquires knowledge from human experts in the form of rules and facts. More advanced technology allows intelligent software to acquire knowledge from different problem domains. The knowledge learned by computer software is more accurate and reliable than that of human experts.

KNOWLEDGE ACQUISITION SUBSYSTEM
A method of checking the knowledge base for incomplete or inconsistent data. The findings are presented to the expert for resolution.

KNOWLEDGE BASE
The collection of rules and facts gathered from experts about a particular subject.

KNOWLEDGE-BASED SYSTEMS
Programs that reason from the information contained in extensive knowledge bases.

KNOWLEDGE DATABASE
The rules and cases used in making decisions.

KNOWLEDGE ENGINEERING
The science of converting expert advice into coded rule sets and constructing a knowledge base. This requires collaboration between the human experts and knowledge engineers familiar with the construction process.

KNOWLEDGE LEVEL
The specific amount of knowledge required by the machine.

LAW OF THE EXCLUDED MIDDLE
An element in classical set theory that relates to a set consisting of either complete membership (true - one) or complete nonmembership (false - zero). An element can be a member of a set and its complement. Because fuzzy sets have part membership characteristics they behave differently in the presence of their complement.

LEARNING RATE
What controls how fast weights change.

LIAR PARADOX
Also called the self-reference paradox. It reduces to literal half-truths in a fuzzy or multivalued logic.

LINEAR SYSTEM
A system in which the whole equals the sum of its parts. All other systems are nonlinear. A linear system can be broken down into small pieces and then the small pieces may be evaluated. The small pieces are then patched to bring them back to the whole system.

LINEAR TIME INVARIANT (LTI) SYSTEM
A system whose structure does not change over time.

LINGUISTIC VARIABLE
The rule formulation language of fuzzy systems. It applies to a fuzzy set or the combination of a fuzzy set and its associated hedges.

LISP
A general purpose artificial intelligence programming language used to design and program parts of expert systems from scratch. LISP is the language most often used for expert systems development. It is a language for writing programs that manipulate symbolic expressions referred to as "list structures," whose elements may be lists, or lists of lists. A list structure is typically denoted by a pair of parentheses enclosing the sequence of its elements. LISP programs can be designed to manipulate more complex programs.

LOGICAL POSITIVISM
The notion that math statements are meaningful and that the meaning of a sentence is in its method of verification.

M-ORDER FUZZY SET
A fuzzy set having fuzzy characteristics.

MAXIMUM
The technique for finding the maximum point in the solution fuzzy set.

METHOD OF DEFUZZIFICATION
The functional relationship between fuzzy regions and the expected value of a set point, describing how the expected value for the final fuzzy state space can be derived.

METHOD OF IMPLICATION
The functional relationship between the degree of truth in related fuzzy regions.

MINIMUM/MAXIMUM RULE OF IMPLICATION
A rule applied to the fuzzy logic process when there is a specification of how fuzzy unions and intersections are performed. If two fuzzy sets are combined with the intersection operator (AND), the ensuing fuzzy space is derived by taking the minimum of the truth functions across the compatible domains. If two fuzzy sets are combined with the union operator (OR), the ensuing fuzzy space is derived by taking the maximum of the truth functions across the compatible domains.

MINIMUM/MAXIMUM RULE OF INFERENCE
A rule applied in a fuzzy system to the way conditional and unconditional fuzzy propositions or assertions are combined.

MODUS PONENS (MP)
A type of implication used in classical fuzzy logic to infer the presence of a consequent state from an antecedent. In fuzzy logic, MP relates to the degree of truth between the premise and consequent.

MODUS TOLLENS (MT)
One type of logical implication process relating to the lack of a premise state, given a negation of the consequent state. In fuzzy logic, MT (like MP) relates to the degree of truth between the premise and the consequent.

MOMENTUM
The tendency of weights inside a unit to change the direction in which they are heading. Each weight recalls whether it is becoming larger or smaller; momentum attempts to keep it moving in the same direction. A network having high momentum responds slowly to new training examples that want to reverse the weights. This is helpful when training examples are ordered by similarity. In a network with low momentum, weights are allowed to oscillate more freely.

MULTIPROCESSING
The actions of several machines.

NATURAL LANGUAGE PROCESSING
A method allowing users to communicate with computers in their native language.

NECESSITY
An inevitable event, measured by the degree to which one event (condition) depends on another.

NEURAL NETWORKS (NNs)
In humans, sets of interconnected neurons and synapses that act like a computer, converting inputs to outputs. In computers, NNs act like "thinking problem-solvers": they are software programs that simulate human intelligence and learn from experience. For example, each

time an NN makes the right decision (as predetermined by a human instructor) in recognizing a number or sequence-of-action pattern, the programmer reinforces the program with a configuration message that is stored; if the decision is wrong, the reinforcement is negative. Gradually, the NN builds experimental knowledge in that subject. Most NNs are mathematical simulations embedded in software.

Mathematical or artificial NNs run on software or on special chips. They are used to promote learning and to recognize patterns, such as the characteristics of a high-risk loan applicant. NNs are sometimes referred to as neurocomputers. The more examples they are given, the better they learn. NNs allow computers to learn from a database and the operator what the right answers are to questions. They learn through exposure to a sample set of inputs paired with the suitable action for each input.

A NN behaves like an associative memory storing associations between inputs and outputs, stimuli, and responses. Theoretically, NNs can model any system if sufficient neurons are used. Practically, they are good at learning separate pattern classes, such as determining the authenticity of a signature on a contract. They can be trained to pick out regularities or patterns in noisy environments. NNs can best be applied when (a) there is no exact model of the underlying process, (b) when there is considerable noise in the environment, or (c) when there is considerable change in the environment. NNs can determine relationships between things not even known to exist. They are ideal for problems when deductive reasoning gives mixed results because of their inductive reasoning ability and their large storage base of historical data that can be appraised for subtle interrelationships. NNs are particularly useful in risk appraisals, economic forecasting, and fraud detection.

NEURONS
Nerve cells. In the human brain there are about 12 billion.

NORMALIZATION
The process used to assure that the maximum truth value of a fuzzy set is one. Fuzzy sets are normalized by readjusting all the truth membership values proportionately around the maximum truth value, which is set to one.

ONTOLOGY
The premise of what exists and why.

ORGANIZATION
A collection of individuals solving problems or conducting functions relative to each other, particularly when the relations and activities are relatively unchanging.

PARADIGM
A general model of something useful for investigating a given thing.

PARALLELISM
The technique in which many different possibilities are examined simultaneously as an efficient way to solve a business problem.

PARALLEL PROCESSING
(1) Machines performing actions simultaneously; (2) A single system making several concurrent decisions. Neural networks use parallel processing to simultaneously evaluate multiple inputs.

PARTIAL SPECIFICATION
A situation that is not fully described so that it is impossible to answer all possible questions about the situation.

PATCHES
Clusters in data. If you learn data clusters, you learn patches, you then learn rules, and you have an adaptive fuzzy system.

PATTERN
A collection of objects, each of which meets a specified criterion, which is the pattern rule for the pattern. Artificial intelligence applications include learning patterns, classifying items, matching items, and describing the environment.

PATTERN EXAMPLES
Objects in a pattern.

PATTERN PERCEPTION
The ability to find a simple, useful description of something, given an initial description that is very complex or of low utility.

PERCEPTUAL ALIASING
A situation in which two different environmental situations result in the same sensory input.

PHYSICAL GROUNDING HYPOTHESIS
The proposition that various behavior modules of an agent interact with the environment to generate complex behavior without the use of centralized models.

PLANNER
A programming language used for artificial intelligence research. The program uses pattern matching to search a database for certain expressions. If those expressions are found, a new expression is added to the database.

PLANNING
Construction and execution of arrangements to solve problems.

POLICY
The logical segment of a fuzzy model. The policy may be functional, operational, or organizational. It is a logical and self-contained unit of knowledge within the overall model.

There are many policies associated with models. In a project risk assessment model, there may be a policy appraising risk based on the physical characteristics of a project (e.g., complexity, organization, time period, funding, complexity, and similarity to previous projects). Another policy may analyze risk by looking at project management factors such as skill required, priority, visibility, and qualifications of the project leader.

PREDICTIVE PRECISION
The extent to which a prediction or proposition of the system is proved or disproved. Imprecise or approximate claims are less verifiable.

PREMISE (PREDICATE)
The conditional part of a rule. In fuzzy modeling, the premise represents a series of fuzzy propositions or assertions connected by AND and OR operators. A premise assertion would be preceded by "if" and end with "then." A fuzzy rule-based system will appraise each premise proposition to find its truth value. These will be combined by

the AND and OR operators to result in a final and total truth for the rule premise.

PROBABILITY

The likelihood of success or failure calculated by putting odds on precise events. This is a quantitative approach for handling uncertainty based on measuring behavior patterns for a selected set of properties in a population. If the a priori likelihood of an event taking place is known, the likelihood of that event taking place in the future can be predicted with a certain degree of confidence. A higher probability means a greater chance of the event occurring. The total of all probabilities must equal 100 percent. Probability theory is based on bivalent logic.

PRODUCT SPACE CLUSTERING

A rule is a cluster in the data. With more clusters, there are more rules.

PROGRAMMING LANGUAGE

A set of sentences (signals) received and stored internally by the computer as a data structure. Different forms of data structure (e.g., numbers, lists, graphs) are used to instruct the computer to conduct actions. The syntax of the programming language is given when the exact forms of sentences and data structures are described. The semantics of the programming language is given when actions are specified for each data structure requiring performance.

PROJECTING

The process of predicting a sequence of states arising from a sequence of actions.

PROLOG

A general-purpose symbolic programming language useful in a variety of artificial intelligence applications.

PROPOSITION

The root mechanism in a fuzzy model referring to a statement of relationships between model variables and one or more fuzzy regions. A series of conditional and unconditional fuzzy propositions or associations is appraised for its extent of truth; all those that have some truth contribute to the final output state of the solution variable set.

PROPOSITIONAL SATISFIABILITY
Finding a model for a formula.

PROTOCYCLING METHODOLOGY
A cyclical method of defining a base system that is refined and extended as different performance measurements are applied.

PROTOTYPE SYSTEM
A system tested by the knowledge engineer and the expert to determine if it makes the same types of inference that the expert would make on representative problems that might be posed by a user. If the response is different from what the expert would do, the explanation subsystem is used to help the development group decide which data inferences are causing the discrepancy—e.g., perhaps the expert needs to provide more information. Prototyping continues until the system is working properly.

PYTHAGOREAN THEOREM
A measurement of how much one fuzzy set contains another.

RAMIFICATION PROBLEM
A challenge to track derived formulas that survive subsequent state transitions.

RANDOMNESS
The characteristic of an event or property such that in a sample there is an equally likely chance of a possible value taking or not taking place. A major characteristic of randomness in an event is the inability to predict an outcome.

REDUCTIO AD ABSURDUM
An assumption shown to result in a contradiction or absurdity that makes it necessary to go back and deny the assumption.

RELATIVE FREQUENCY (SUCCESS RATIO)
Successes divided by trials. Every trial is either a success or failure.

REPRESENTATION
Finding programming languages that state pattern rules.

Glossary

ROBOT
A mechanical intelligence able to operate in the real world environment. A robot should have sensing, reasoning, and perception abilities so that it can detect patterns in a prescribed environment. It should also have a way to act upon its environment.

RULE-BASED REASONING
Use of a set of user preconditions to appraise conditions in the outside environment.

RULE-BASED EXPERT SYSTEM
An expert system contains a set of deductive knowledge production rules. Each rule has an if-then clause. Users provide facts or statements so that production rules can be triggered and the conclusion can be generated.

RULES
Knowledge statements relating the compatibility of fuzzy premise propositions to the compatibility of one or more consequent fuzzy spaces.

SCALABLE MONOTONIC CHAINING
Mapping the risk specified in individual rules to an intermediate risk-measuring fuzzy set to arrive at a final risk value.

SCHEMA
A term used in genetic algorithms meaning form or figure, applying to a representation of patterns.

SEARCH SPACE
A collection of candidate solutions to a problem, including determining the distance between candidate solutions.

SELF-ORGANIZING SYSTEM (SOS)
A collection of machines able to solve problems by forming into organizations. In an SOS, there is an environment or collection of problems to be solved and patterns to be recognized.

SEMANTIC INFORMATION
Sentences in a program language that have meaning to those who use the languages.

SERIAL PROCESSING
A system making only one decision at a time. Expert systems use serial processing. For example, an expert system processes a series of if-then rules, matching each "if" with its "then" rule until a final result is reached.

SET
A system or collection of things.

SET THEORY
A review of sets or classes of objects. In math, the basic unit is the set. In logic, the basic unit is the symbol. The foundations of math are set and logic theory.

SIMULATION
A computer program duplicating the behavior of something, such as the actions of a human expert in solving a business problem.

SITUATION CALCULUS
The predicate-calculus formalization of states, actions, and the impact of effects of actions on states.

SITUATION-SPACE MODEL
An approach to solving a problem in which there is an initial situation, a set of possible situations, and a specification of possible actions, along with enumeration how the various situations can be generated from each other by different actions and highlighting of a final suggested situation or goal. The statement of the problem may also include outlining situations to be avoided. A solution to a situation-space problem is any sequence of actions that progresses from the original situation to the desired situation while avoiding undesirable situations.

SORITES
A logical chain of statements in the form of "if A, then B; if B, than C; etc."

SORITES PARADOX
A paradox consisting of a chain of bivalent if-then statements.

Glossary

STATIC EVALUATION FUNCTION
A method of estimating the value of a node that is not dependent on the value of the successors to that node.

SUBSETHOOD
The extent to which one set includes another set (the containment value). In classical theory, a set has the subsets all or none. In fuzzy logic, it is a matter of degree. Thus, the subsethood or containment value can have any value between 0 percent and 100 percent.

SUPERVISED LEARNING
Use of a teacher to manage the learning process in neural networks by furnishing correct output values. If there is no supervised learning and no teacher, the learning system learns on its own via environmental feedback on its performance.

SUPPORT SET
The component of a fuzzy region or fuzzy set expressed with a strong alpha-cut threshold. It is the area participating in the fuzzy implication and inference process.

SYMBOL LEVEL
How knowledge is represented in symbolic structures, such as lists written in the programming language LISP and specified operations on these structures.

SYSTEM INFERENCE
A solution to problems that take many varying forms of representation, all of them theoretically equivalent, that the computer finds to be more efficient than the others. The inference proposed as a solution will usually be a finite description of a function, relation, or mathematical theory.

TERM SET
The collection of fuzzy sets associated with a variable.

THREE-VALUED LOGIC
A logic operating with statements that are true, false, or indeterminate.

TOP-DOWN
A system where human experts write sophisticated and complex rules.

TOP-DOWN DESIGN
A symbol-processing method that begins at the knowledge level and proceeds downward through the symbol.

TRAINING EPOCH
In the neural network learning process, the full set of cycles of training inputs. In most cases, many training epochs are required before a neural network learns how to classify a particular set of training inputs efficiently.

TRAINING LEARNING UNIT (TLU)
The unit being trained by a neural network. If a TLU is used to compute an action, its inputs must be numeric so that a weighted sum may be derived. A TLU is trained by modifying its variable weights. A TLU may be defined by weights and threshold.

TRUTH FUNCTION
The degree to which a variable's value is compatible with the fuzzy set (the degree of membership). The function is a situation between no membership (zero) and full membership (one). The degree of membership is drawn from the truth function of the fuzzy set.

TURING'S TEST
An experiment to determine if a machine possesses intelligence on a human level.

UNCERTAINTY PRINCIPLE
The principle that as one option grows with more certainty, the other option has less certainty.

UNDECIDABILITY
The extent to which the state of a control variable cannot be ascertained from the fuzzy space context.

UNION
The disjunction of two fuzzy sets.

UNIVERSAL (ACTION) PLAN
The specification of an action for every possible state in a reactive program.

USER INTERFACE
The explanatory features, on-line help facilities, debugging tools, modifications systems, and other tools designed to help the user effectively use the system.

VIRTUAL (ARTIFICIAL) REALITY
Cyberspace. Virtual reality relates to tricking the senses of the mind in the computer world.

ZADEH'S FUZZY ALGEBRA
Strict rules associated with the combination of fuzzy intersection and union. The union AND relates to the maximum of two fuzzy sets while the intersection OR refers to the minimum of two fuzzy sets.

Index

A

AASYST. *See* Applied sentencing systems.
Access Elf, 183
Accounts payable, 101
   See also Artificial intelligence in accounting; Expert system applications.
Accounts receivable, 94, 111-117
   Credit granting, 111
   Detection of credit fraud, 114
   Monitoring credit and writing off bad debt, 116
   See also Artificial intelligence in accounting; Expert system applications; Neural networks in accounting.
ACE, 29
Acquire, 182
Acquired Intelligence Industries, 182
Adaptive, 189
Adaptive fuzzy associative memory (AFAM), 189
Adaptive fuzzy systems (AFS), 189
Advice on trading, 56-57
AFAM. *See* Adaptive fuzzy associative memory.
AFS. *See* Adaptive fuzzy systems.
AICPA Statement, 127, 129
Aiken, N.W., 154, 168

AI Ware, 181
Alpha-level, 190
Ambiguity, 190
"An Empirical Test of Financial Ratio Analysis for Small Business Failure Prediction," 47
Antecedent, 190
APM, 166
Applications in finance. *See* Banking; Insurance; Portfolio management; Trading advice.
Applications of expert systems, 28-36
   Accounting systems, 30
   Analysis of credit and loan applications, 32
   Capital expenditure planning, 32
   Finance, 34-36
   Global financial market, 36
   Marketing, 33
   Other applications, 37-40
   Strategic planning, 33
   See also Artificial intelligence in business finance; *CoverStory; Intelligent Scheduling and Information System (ISIS-II); CARGEX-Cargo Expert System; NCR Corporation ES; ACE; DELTA; XCON; Authorizer's Assistant (AA); Watchdog Investment Monitoring System; Escape; Plan Power; GURU; Peat Marwick ES; XSEL; Optex; PC-Based Performance Analyst System.*

223

Index

Applied sentencing systems, 138
   See also Legal applications.
Approximate reasoning, 190
Approximation, 190
Arbitrage. *See* Global financial markets.
Artificial intelligence and expert systems, 23-40
Artificial intelligence applications, 2
   Accounting applications, 4
   Business applications, 3
   Finance applications, 5
   Marketing applications, 6
   See also Artificial intelligence in business finance; Artificial intelligence in accounting.
Artificial intelligence, 191
Artificial intelligence in accounting, 93-133
   Expert system applications, 93-107
   Neural networks in accounting, 107-133
Artificial intelligence in business finance, 49-91
   Advice on trading, 56
   Banking, 50
   Global financial markets, 57
   Insurance, 51
   Portfolio management, 53
Artificial intelligence in marketing, 149-169
   Communication, 165
   Forecasting, 160
   Inventory management, 164
   Making decisions in uncertainty
   Marketing services, 159
   Selling, 153
   Trend analysis, 165
   Using AI at the design stage, 151
Artificial intelligence software, 181-187
Artificial languages, 191
Assessing the going concern assumption, 130-132
   Commercial software packages and services, 132
   Follow-up and monitoring, 132
   Identifying and selecting data, 131
   Planning the NN application, 131
   Retraining the NN, 132
   Training the NN, 131
   Using the NN in the going concern evaluation, 131
   See also Artificial intelligence in accounting; Expert system applications.
Assertions, 191
Assignment, 191
Associative memory, 191

Auditing, 102, 127-133
   Assessing the going concern assumption, 130
   Fraud detection, 129
   Internal control assessments, 127
   Operational auditing, 132
   See also Neural networks in accounting.
Audit Standard No. 78, 127
Audit Standard No. 82, 129
Auditor, 113
Authorizer's Assistant (AA), 29, 113
Automatic programming, 38, 192
   See also Branches of AI.

B

Backward chaining, 15, 192
Baets, W., 169
Bank check fraud, 68-70
   Commercial software packages and services, 69
   Follow-up and monitoring, 69
   Identifying and selecting data, 68
   Planning the NN, 68
   Retraining the NN, 69
   Training the NN application, 69
   Using the NN to detect fraud, 69
   See also Neural networks in business finance.
Banking, 34
   Bank check fraud, 68
   Credit and debit fraud, 65
   Credit granting, 50, 61
   Credit monitoring, 51
   Fiduciary responsibilities, 70
   See also Neural networks in business financing.
Bankruptcy prediction models, 44
BANKSTRAT, 159
Barnes, J.H., 154, 168
Bayesian networks. *See* Branches of AI.
Bayes theorem, 192
Belief system, 192
Ben, Akiva, M., 161, 168
BioCamp Systems, Inc., 182
Bit value, 192
Bivalence, 192
Bivalent logic, 192
Bot. 192
Bots and agents, 15
Bottom-up, 193
BPM. *See* Business processing modeling.
Brain Maker Professional, 183

Index

Branches of AI, 14
Budgeting, 124-125
   Commercial software packages and services, 125
   Follow-up and monitoring, 125
   Identifying and selecting data, 124
   Planning the NN application, 124
   Retraining the NN, 125
   Training the NN to support budgeting, 125
   Using the NN to detect excessive budgetary hedging, 125
   See also Neural networks in accounting.
Business Advisor, 181
Business neural network applications, 42
Business processing modeling (BPM), 193
Business process reengineering, 176
   See also Expert systems and neural networks in manufacturing.

C

CAD/CHEM, 181
California Scientific Software, 183
Capital Investment Expert System, 187
Capstone Application Decision Processing System, 184
Capstone Decision Manager, 74
Capstone Online, 65
Capstone™ Decision Manager, 64
Cardholder Risk Identification Service, 68
CARGEX-Cargo Expert System, 28
Carroll, J. Douglass, 152, 168
Cartesian product, 193
Case-based expert system, 193
Case-based reasoning, 193
Cash management, 94.
   See also Artificial intelligence in accounting; Expert system applications.
C/C, 14
Centroid method, 193
Chaos, 194
Chromosome, 194
CI. *See* Compatibility index.
Clusters, 194
CM. *See* Composite maximum.
Cognitive structure, 194
Communication, 165
   See also Artificial intelligence in marketing.
COMP, 164
COMPADVISOR™, 65
Compatibility, 194
Compatibility index (CI), 194

Compensatory operations, 195
Complement, 195
Complex Bill Service and Mail-in Bill Review Service, 75
Composite mass, 195
Composite maximum (CM), 195
Conjunction, 196
Conjunctive normal form, 196
Connectionism, 196
Consequent, 196
Consistency principle, 196
Consulting systems, 196
Consumer behavior, 160
   See also Artificial intelligence in marketing.
Containment, 196
Continuous variable, 196
Contrast, 196
Correlation encoding, 196
Correlation minimum rule, 196
Correlation product rule, 197
CoverStory, 28
Credit and debit card fraud, 65-68
   Commercial software packages and services, 67
   Follow-up and monitoring, 67
   Identifying and selecting data, 66
   Planning the NN, 65
   Retraining the NN, 66
   Training the NN application, 66
   Using the NN for fraud detection, 66
   See also Neural networks in business finance.
Credit granting, 111
   Commercial software packages and services, 113
   Follow-up and monitoring, 113
   Identifying and selecting data, 112
   Planning the NN, 112
   Retraining the NN, 113
   Training the NN application, 112
   Using the NN in granting credit, 113
   See also Accounts receivable.
Credit Intelligence Solutions, 67, 115
Crisp set, 197
Critics. *See* Advice on trading.
Curry, B., 168

D

Data analysis, 46
Database management system, 197

Database management system. *See* How expert systems work.
Datafeed Toolbox, 71, 109
Dataflow, 197
Data set ratio definitions, 46
Davies, F., 168
Decision support system (DDS), 23, 197
Decomposition method, 197
Deductive rule-learning, 197
Deferred Income Taxes, 103.
    See also Artificial intelligence in accounting; Expert system applications.
Defuzzification, 197
Delphi method, 198
DELTA, 29
Del Valle, Christina, 156, 168
Dempster-Shafer theory, 198
Detection of credit fraud, 114
    Commercial software packages and services, 115
    Follow-up and monitoring, 115
    Identifying and selecting data, 114
    Planning the NN, 114
    Retraining the NN, 115
    Training the NN application, 114
    Using the NN in granting credit, 115
    See also Accounts receivable.
Difference operators, 198
Digital signal processing, 198
Dilution, 198
Direct marketing, 155-159
    The business consumer, 157
    The final consumer, 155
    See also Artificial intelligence in marketing.
Disjoint fuzzy space, 198
Disjunction, 198
Distributed artificial intelligence, 198
Domain, 198
Domain Database, 199
Domain database. *See* How expert systems work.
Domestic violence, 140
    See also Legal applications.
DSS. *See* Decision support system.
Dynamic decision network, 199
Dynamical system, 199

E

Eagle, 90
Earnings per share, 103.
    See also Artificial intelligence in accounting; Expert system applications.

Edge, 199
Edmonds, B., 169
Elementhood, 199
Elf software company, 183
Eliza, 12-13
Emergent behavior, 199
Employee expense reimbursement, 121-122
    See also Neural networks in accounting.
Englis, B.G., 156, 168
Entropy, 199
ES. *See* Expert system.
Escape, 29
Establishing reserves, 75-77
    Commercial software packages and services, 77
    Follow-up and monitoring, 77
    Identifying and selecting data, 76
    Planning the NN application, 76
    Retraining the NN, 77
    Training the NN application, 75
    Using the NN for the reserving function, 76
    See also Neural networks in business finance.
ES Tools, 26-28.
    See also Languages for expert system development; Expert system shells and products; Popular ES shells.
Evolutionary system, 199
Expected value, 199
Expert Series, 186
Expert system (ES), 199
Expert system applications, 93-107
    Accounts receivable, 94
    Accounts payable, 101
    Auditing, 102
    Cash management, 94
    Deferred income taxes, 103
    Earnings per share, 103
    Income taxes, 106
    Inventory, 96
    Long-term debt, 102
    Payroll, 102
    Prepaid expenses, 96
    Securities, 94
Expert system shell, 200
Expert system shells and products, 2, 17
Expert systems and neural networks in manufacturing, 171-179
    Business process reengineering, 176
    Example one, 172
    Example two, 173
    Production and operations, 174
    Robotics, 177

Index

Expert systems in detail, 24-25
   Benefits, 36
   Disadvantages, 37
Explanation subsystem, 201

F

Falcon™, 67
Falcon Payment Card Fraud Detection System, 183
Falcon™ *Debit*, 67
Falcon™ *Model 2000*, 67
FAM. *See* Fuzzy associative memory.
Fat theorem, 201
FCM. *See* Fuzzy cognitive map.
Fenn, J. 168
Fiduciary responsibilities, 70-71
   Commercial software packages and services, 71
   Follow-up and monitoring, 71
   Identifying and selecting data, 70
   Planning the NN, 70
   Retraining the NN, 71
   Training the NN to meet fiduciary responsibilities, 70
   Using the NN in fiduciary functions, 71
   See also Neural networks in business finance.
Finance. *See* Neural networks in business finance.
Financial Advisor. See Popular ES shells.
"Financial Ratios as Predictors of Failure," 47
"Financial Ratios, Discriminant Analysis and the Prediction of Corporate Bankruptcy," 47
Financial Solutions, 67
Financial statement advice. *See* Global financial markets.
Financial Systems, 117
FINDER, 136
   See also Legal applications.
Fish, K.E., 154, 168
Fit value, 202
Fixed point attractor, 202
Flacon Cheque™, 69
Flagging potential fraud, 88-90
   Commercial software packages and services, 90
   Follow-up and monitoring, 90
   Identifying and selecting data, 89
   Planning the NN application, 89
   Retraining the NN, 89
   Training the NN, 89
   Using the NN for fraud detection, 89
   See also Investment banking.
Flagging the potential for fraud, 85-87
   Commercial software packages and services, 87
   Follow-up and monitoring, 87
   Identifying and selecting data, 86
   Planning the NN application, 86
   Retraining the NN, 86
   Training the NN, 86
   Using the NN to detect fraud by developers, 86
   See also Real estate brokers.
Forecasting, 160
   Consumer behavior, 160
   The actual forecast, 161
   See also Artificial intelligence in marketing.
Forward chaining, 202
Fraud detection, 129-130
   Commercial software packages, 130
   Follow-up and monitoring, 130
   Identifying and selecting data, 130
   Planning the NN, 129
   Retraining the NN, 130
   Training the NN, 130
   Using the NN to predict control risk, 128
   See also Artificial intelligence in accounting; Expert system applications.
Fuzziness, 202
Fuzzy associative memory (FAM), 202
Fuzzy (cloudy, vague) logic, 203
Fuzzy cognitive map (FCM), 202
Fuzzy entropy, 203
Fuzzy knowledge engineering, 203
Fuzzy logic, 16, 37
Fuzzy model, 203
Fuzzy numbers, 203
Fuzzy operators, 203
Fuzzy principle, 204
Fuzzy reasoning, 204
Fuzzy rules, 204
Fuzzy sets, 204
Fuzzy space, 204
Fuzzy system, 204

G

Games of strategy, 205
GAs. *See* Genetic algorithms.
Gemini Application Fraud Detector, 67
Gemini Verify Score, 64
Generalized delta rule, 205
Generations process, 205
Generator, 9, 205
Generator TFV, 9, 10

Genetic algorithms (GAs), 9, 205
Genetic programming, (GP), 206
Global financial markets, 36, 57-60
   Arbitrage, 59
   Financial statement advice, 58
   Hedges, 59
   Twenty-four-hour trading programs, 58
Goodman, S. 168
GP. *See* Genetic programming.
Granting credit, 61-65
   Commercial software packages and services, 64
   Follow-up and monitoring, 64
   Identifying and selecting data, 62
   Planning the NN for granting credit, 61
   Retraining the NN, 64
   Training the NN application, 63
   Using the NN, 63
   See also Neural networks in business finance.
Green, Paul E., 152
GURU, 29

H

Hedge, 206
Hedges. *See* Global financial markets.
Heuristic search, 206
Hodgdon, P., 168
How expert systems work, 25-26
   Relationships, 26
HyperText Markup Language (HTML), 18

I

i24™, 67
IBM, 184
ICaR, 140
   See also Legal applications.
Implication, 207
Imprecision, 207
Income taxes, 106
   Tax compliance, 106
   See also Artificial intelligence in accounting; Expert system applications.
Inductive rule-learning, 207
Inference engine, 207
   See also How expert systems work.
Inheritance advisor, The 140
   See also Legal applications.
Innovative, 207
Insolvency prediction, 43
Instant tea, 17

Insurance, 34
   Claims processing, 52
   Insurance claim fraud, 74
   Establishing reserves, 75
   Prescreening applications, 72
   Reserving, 53
   Underwriting, 52, 72
   See also Neural networks in business finance.
Insurance claim fraud, 74-75
   Commercial software packages and services, 75
   Follow-up and monitoring, 75
   Identifying and selecting data, 75
   Planning the NN application, 74
   Retraining the NN, 75
   Training the NN application, 75
   Using the NN for fraud detection, 75
   See also Neural networks in business finance.
Integrated Reasoning Shell for OS/2 Release 3, 184
Intelligent agents, 39-40, 207
Intelligent computer-aided instruction (ICAI), 40
Intelligent Scheduling and Information System (ISIS-11), 28
Intensification, 207
Internal control assessments, 127-129
   Commercial software packages and services, 129
   Follow-up and monitoring, 129
   Identifying and selecting data, 128
   Planning the NN, 128
   Retraining the NN, 129
   Training the NN, 128
   Using the NN to predict control risk, 128
   See also Artificial intelligence in accounting; Expert system applications.
Internal Operations Risk Analysis Software, 129
Intersection, 207
Intrinsic fuzziness, 208
"Introducing Recursive Partitioning for Financial Classifications: The Case of Financial Distress," 48
Invention, 208
Inventory, 96
   ABC Allocation methods, 98
   Manufacturing conversion of inventory, 97
   Standard cost accounting system, 99
   See also Artificial intelligence in accounting; Expert system applications.

Index

Inventory accounting, 117-121
  Inventory control, 118
  Inventory valuation, 119
  See also Neural networks in accounting.
Inventory control, 118
  Commercial software packages and services, 119
  Follow-up and monitoring, 119
  Identifying and selecting data, 118
  Planning the NN application, 118
  Retraining the NN, 119
  Training the NN, 119
  Using the NN to detect inventory errors or fraud, 119
  See also Neural networks in accounting.
Inventory management, 164
  See also Artificial intelligence in marketing.
Inventory valuation, 119
  Commercial software packages and services, 121
  Follow-up and monitoring, 121
  Identifying and selecting data, 120
  Planning the NN application, 120
  Retraining the NN, 121
  Training the NN, 120
  Using the NN, 120
  See also Neural networks in accounting.
Investment banking, 87-90
  Flagging potential fraud, 88
  Setting the selling price for securities, 87
  See also Neural networks in business finance.

J

Java expert system shell (Jess), 17
Journal of Advertising, 156
Journal of Finance, 47
Journal of Financial and Quantitative Analysis, 47
Journal of Marketing Research, 152

K

Kalman filter, 208
Kinnear, T.C., 168
Kleege, S., 168
Knoowledgepro. See Popular ES shells.
Knowledge acquisition facility, 208
Knowledge acquisition facility. *See* How expert systems work.
Knowledge acquisition subsystem, 208

Knowledge base, 208
Knowledge-based systems, 208
Knowledge database, 209
  See also How expert systems work.
Knowledge engineering, 209
Knowledge level, 209
Knowledge Tool for As/400, 184
Kotler, Philip, 149, 168
Kramer, B., 167, 168

L

Languages for expert system development, 26
Law of the excluded middle, 209
Learning rate, 209
Legal applications, 135-147
  Applied sentencing systems, 138
  Domestic violence, 140
  FINDER, 136
  Inheritance advisor, 140
  Legal information retrieval system, 139
  Negligence, 140
  Nervous shock advisor, 139
  Sample consultation hypothetical for NSA, 142
  SHYSTER, 137
Legal information retrieval system, 139
  See also Legal applications.
Leonardo. See Popular ES shells.
Lerman, S.R., 161, 168
Level 5 Object Professional Release 3.0, 184
Level 5 Quest, 184
Level 5 Research, 184
Liar paradox, 209
Liew, Chung J., 171
Linear system, 209
Linear time invariant (LTI) system, 209
Linguistic variable, 209
LISP, 14, 210
Loan Performance Score, 85
Logical positivism, 210
Long-term debt, 102
  See also Artificial intelligence in accounting; Expert system applications.
LTI. *See* Linear time invariant.

M

Machine Learning, *See* Branches of AI.
Making decisions in uncertainty
  See also Artificial intelligence in marketing.
Malhotra, Naresh K., 152, 169

Manufacturing. *See* Expert systems and neural networks in manufacturing.
Marketing services, 159
   See also Artificial intelligence in marketing.
MAT-LABR, 71
Maximum, 210
MDSCAL 5M, 166
Method of defuzzification, 210
Method of implication, 210
Minimum/maximum rule of implication, 210
Minimum/maximum rule of inference, 211
MIRAR, 77
Model Ware, 187
ModelWare Professional, 186
Modus ponens (M)), 211
Modus tollens (MT), 211
Momentum, 211
Monitoring credit and writing off bad debt, 116
   Commercial software packages and services, 117
   Follow-up and monitoring, 117
   Identifying and selecting data, 116
   Planning the NN, 116
   Retraining the NN, 117
   Training the NN application, 117
   Using the NN in granting credit, 117
   See also Accounts receivable.
Montgomery, D., 169
M-order fuzzy set, 210
Morrison, Ann M., 151, 169
Moss, S., 169
Mountinho, L., 168
MP. *See* Modus ponens.
MT. *See* Modus tollens.
Multi Logic, 185
Multi Logic Exsys Professional, 185
Multiprocessing, 211

N

Natural language. *See* Branches of AI.
Natural language processing (NLP), 39, 211
NCR Corporation ES, 28
Necessity, 211
Negligence, 140
   See also Legal applications.
Nervous shock advisor, 139
   See also Legal applications.
Neural Network Bankruptcy Prediction Program, 45
Neural Network Utility Product Family, 184

Neural networks in accounting, 107-133
   Accounts receivable, 111-117
   Auditing, 127-133
   Budgeting, 124-125
   Employee expense reimbursement, 121-122
   Inventory accounting, 117-121
   Payroll control, 122-124
   Performance measurement, 125-127
   Securities, 107-111
Neural networks in business finance, 60-91
   Banking, 60-71
   Insurance companies, 71-77
   Investment banking, 87-90
   Portfolio management, 80-81
   Real estate brokers, 81-87
   Retail securities brokers, 90-91
   Trading advisory services, 77-80
Neural networks (NNs), 41-48, 211
Neural Systems, Inc., 185
Neural Ware Inc., 185
Neural Works Predict, 185
Neuro Genetic and Trade, 182
Neurons, 212
NeuroShell Toolbox, 109
Nexcore[TM], 65
NNs. *See* Neural networks.
Normalization, 212
N-Train: Neural Network System, 186

O

OCM24[R], 67
Ontology, 213
Operational auditing, 132-133
   Control procedures, 133
   Data-mining, 132
   Process enhancement, 132
   See also Artificial intelligence in accounting; Expert system applications.
Optex, 29
Organization, 213

P

Paradigm, 213
Parallelism, 213
Parallel processing, 213
Partial specification, 213
Patches, 213
Pattern, 213
Pattern examples, 213
Pattern perception, 213

Index

Payroll, 102
   See also Artificial intelligence in accounting;
      Expert system applications.
Payroll control, 122-124
   Commercial software packages and services, 124
   Follow-up and monitoring, 123
   Identifying and selecting data, 123
   Planning the NN application, 123
   Retraining the NN, 123
   Training the NN, 124
   Using the NN to detect payroll error or
      fraud, 123
   See also Neural networks in accounting.
PC AI Magazine, 48
PC-Based Performance Analyst System, 30
Peat Marwick ES, 29
Perceptual aliasing, 214
Performance measurement, 125-127
   Commercial software packages and services, 127
   Follow-up and monitoring, 127
   Identifying and selecting data, 126
   Planning the NN, 126
   Retraining the NN, 127
   Training the NN to detect performance
      inconsistencies , 126
   Using the NN, 127
   See also Neural networks in accounting.
Personal Consultant (PC). See Popular ES
   shells.
Physical grounding hypothesis, 214
Plan Power, 29
Planner, 214
Planning, 214
Plouff, G., 168
Policy, 214
Popular ES shells, 28
Portfolio management, 35, 53, 80-81
   Commercial software packages and services, 81
   Consistent application of constraints, 54
   Follow-up and monitoring, 81
   Hedge advisor, 55
   Identifying and selecting data, 81
   Planning the NN application, 81
   Retraining the NN, 81
   Security selection, 54
   Training the NN for portfolio management, 81
   Using the NN for portfolio management, 81
   See also Neural networks in business
      finance.

Predicting the selling price, 82-83
   Commercial software and services, 83
   Follow-up and monitoring, 83
   Identifying and selecting data, 83
   Planning the NN application, 82
   Retraining the NN, 83
   Training the NN to predict selling price, 83
   Using the NN in listing properties, 83
   See also Real estate brokers.
Predictive precision, 214
Predictive Software Solutions, 77
PREFMAP, 166
Premise (predicate), 214
Prepaid expenses, 96
   See also Artificial intelligence in accounting;
      Expert system applications.
Prequalifying buyers and developers, 83-85
   Commercial software and services, 85
   Follow-up and monitoring, 85
   Identifying and selecting data, 84
   Planning the NN, 84
   Retraining the NN, 85
   Training the NN application, 85
   Using the NN, 85
   See also Real estate brokers.
Prescreening applications, 72-74
   Commercial software and services, 74
   Follow-up and monitoring, 74
   Identifying and selecting data, 73
   Planning the prescreening NN, 72
   Retraining the NN, 74
   Training the NN policy-screening
      applications, 73
   Using the NN for prescreening applications,
      73
PRISM CardAlert, 130
Probability, 215
Process Advisor, 182
Processing, 13
Production and operations, 174
    See also Expert systems and neural networks
      in manufacturing.
Product space clustering, 215
Profit 2000™, 109
ProfitMax Bankruptcy, 132
ProfitMax® Margin Manager, 91
Programming for AI, 14
Programming language, 215
Projecting, 215
Prolog, 14, 215
Proposition, 215
Propositional satisfiability, 216

Protocycling methodology, 216
Prototype system, 216
ProviderCompare[R], 77
Pythagorean theorem, 216

R

Ramification problem, 216
Randomness, 216
Real estate brokers, 81-87
   Flagging the potential for fraud, 85
   Predicting the selling price, 82
   Prequalifying buyers and developers, 83
   See also Neural networks in business finance.
Real-time data feed. *See* Advice on trading.
Reductio ad absurdum, 216
Relative frequency (success ratio), 216
Representation, 216
Retail securities brokers, 90-91
   Commercial software packages and services, 91
   Follow-up and monitoring, 91
   Identifying and selecting data, 90
   Planning the NN application, 90
   Retraining the NN, 91
   Training the NN for margin account management, 91
   Using the NN to manage margin accounts, 91
   See also Neural networks in business finance.
Risk Auditor, 130
Robot, 15, 217
Robotics, 177
   See also Expert systems and neural networks in manufacturing.
Rule-based reasoning, 217
Rules, 217

S

Sample consultation hypothetical for NSA, 142
   See also Legal applications.
Scalable monotonic chaining, 217
Schema, 217
Scientific Consultant Services, Inc., 186
Search space, 217
Securities, 94, 107-111
   Selecting securities, 108
   Trading securities, 109

Securities, *Cont'd*
   See also Artificial intelligence in accounting; Expert system applications; Neural networks in accounting.
Selecting securities, 108-109
   Commercial software packages and services, 109
   Follow-up and monitoring, 109
   Identifying and selecting data, 109
   Planning the NN application, 108
   Retraining the NN, 109
   Training the NN for selecting securities, 108
   Using the NN in selecting securities, 108
   See also Neural networks in accounting.
Self-organizing system (SOS), 217
Selling, 153
   Direct marketing, 155
   The Internet, 153
   See also Artificial intelligence in marketing.
Semantic information, 217
Serial processing, 218
Set theory, 218
Setting the selling price for securities, 87-88
   Commercial software packages, 88
   Follow-up and monitoring, 88
   Identifying and selecting data, 88
   Planning the NN application, 88
   Retraining the NN, 88
   Training the NN to predict selling price, 88
   Using the NN to predict the optimal selling price of a security, 88
   See also Investment banking.
SHYSTER, 137
   See also Legal applications.
Simulation, 218
Situation calculus, 218
Situation-space model, 218
Solomon, M., 156, 168
Sorites paradox, 218
Sorites, 218
SOS. *See* Self-organizing system.
Speech recognition. *See* Branches of AI.
SPEEDMARK, 164
Static evaluation function, 219
Sterling Wentworth Corp., 186
Subsethood, 219
Subsymbolic processing, 8
Supervised learning, 219
Support set, 219
Swinnen, G., 169
Symbol level, 219
Symbolic processing, 6
System inference, 219

Index

T

Taylor, J.R., 168
Teranet, 186
Term set, 219
Texas Instruments Corp., 187
The actual forecast, 161-164
   Data analysis, 163
   Data collection, 163
   Output generation, 163
The Edge™, 64, 113
The Internet, 153
   See also Artificial intelligence in marketing.
Theory development, 162
   See also Artificial intelligence in marketing.
Thomas, C., 172
Three-valued logic, 219
TICOM, 129
TLU. *See* Training learning unit.
Top-down, 220
Top-down design, 220
Trading advice, 35, 78-79
   Commercial software packages and services, 79
   Follow-up and monitoring, 79
   Identifying and selecting data, 78
   Planning the NN to provide trading advice, 78
   Retraining the NN, 79
   Training the NN application, 78
   Using the NN to provide trading advice, 78
   See also Trading advisory services.
Trading advisory services, 77-81
   Trading advice, 35, 78
   Trading rules and rule generators, 79
   See also Neural networks in business finance.
Trading rules and rule generators, 79-80
   Commercial software packages and services, 80
   Follow-up and monitoring, 80
   Identifying and selecting data, 79
   Planning the NN application, 79
   Retraining the NN, 80
   Training the NN for trading rules and rule generators, 80
   Using the NN to formulate trading rules and rule generators, 80
   See also Trading advisory services.
Trading securities, 109-111
   Commercial software packages, 111
   Follow-up and monitoring, 111
   Identifying and selecting data for trading securities, 110

Trading securities, *Cont'd*
   Planning the NN application, 110
   Retraining the NN, 110
   Using the NN in trading securities, 110
   See also Neural networks in accounting.
Trading Simulator, 186
Trading Solutions, 109
Training epoch, 220
Training learning unit (TLU), 220
TRANS24^R, 67
Trend analysis, 165
   See also Artificial intelligence in marketing.
Triant Technologies, 187
Truth function, 220
Tucker, M.J., 169
Turing's test, 10-11, 220
   Can computers think, 20
   Questions, 22
Twenty-four-hour trading programs. *See* Global financial markets.

U

Uncertainty principle, 220
Undecidability, 220
Underwriting, 52, 72
Uniform Commercial Code, 139
Union, 220
Unit Support System, 113
Universal (action) plan, 221
User interface, 221
User interface. *See* How expert systems work.
Using AI at the design stage, 151-153
   See also Artificial intelligence in marketing.

V

Vagueness, 221
Vanhoof, K., 169
Vass, L., 169
Venugopal, V., 169
Virtual (artificial) reality, 221
Visual pattern recognition. *See* Branches of AI.

W

Watchdog Investment Monitoring System, 29
Web sites. *See* Fuzzy logic; Instant tea; Java expert system shell; Turing's test.
What is a machine? 6-9.
   See also Subsymbolic processing; Symbolic processing.

WYSIWYG, 17

X

XCON, 29
XSEL, 29

Z

Zadeh's fuzzy algebra, 221